EARTHQUAKES and
FAULTS in
SAN DIEGO COUNTY

PHILIP KERN

Professor of Geology
San Diego State University

The Pickle Press
San Diego

ON THE COVERS

The map on the front shows the major active faults that produce earthquakes in San Diego County. The stars are epicenters of strong earthquakes that have struck the County since 1800.

On the back is a graphic representation of the strong earthquake history of San Diego County. It shows the shaking intensities (described in chapter VI) of all the well-documented strong quakes that have shaken the County since 1850.

Designed and produced by the author.

A Pickle Press book

ISBN 0-9622845-0-5

ACKNOWLEDGEMENTS

Many people have contributed in a variety of ways to the completion of this book. Much of what is known about earthquake hazard in San Diego County has come from the research of my colleague, Professor Tom Rockwell, and the many students who have collaborated with him on investigations of the San Jacinto, Elsinore, Rose Canyon, and Agua Blanca and offshore fault systems. Tom has given me many hours of his time in discussion of the geologic and seismologic character of these fault zones. He also has provided several of the illustrations used here.

Others who have made substantial contributions to my understanding of this topic include Duncan Agnew, Research Seismologist at Scripps Institution of Oceanography; Steve Day, Eckis Professor of Seismology at San Diego State University; Monte Marshall, Professor of Geology at SDSU; and Mike Reichle, Seismologist with the California Division of Mines and Geology. Professor Marshall read the manuscript and offered many useful suggestions for its improvement, and he also provided several of the illustrations. Additional illustrations were made available by another colleague, Professor Rick Miller.

I take this opportunity to thank those readers of the original edition who suggested changes that may have been incorporated here. Also to be acknowledged are all my colleagues in Geological Sciences who have contributed in various ways to the body of knowledge on which this work is based. Among the many community leaders whose efforts on behalf of earthquake preparedness have helped to inspire this work, the principal individuals are Jan Decker and Dan Eberle at the County Office of Disaster Preparedness; Supervisor Susan Golding, who created the Committee on Earthquake Preparedness; and the members of that committee and its several subcommittees. The efforts of all these people have put San Diego County into a leadership position in California earthquake awareness programs.

The visitor center that is planned for Tecolote Canyon Natural Reserve, and especially its use for earthquake awareness education, will be the reward of years of effort by Karl Anderson and fellow members of the Tecolote Canyon Citizens Advisory Committee. In this endeavor they have been supported by San Diego City Parks and Recreation officials -- especially Pete Jungers, Don Prisby, and Director George Loveland, and by sixth district city councilman Bruce Henderson. The Rose Canyon fault zone model that will serve as the centerpiece of the earthquake hazard exhibit there was built by SDSU geology student Bruce Bailey and was made possible by grants from the Mission Valley and San Diego Exchange Clubs, thanks to member Dave Dunn.

Once again printing consultant Ed Zimmerly has been an indispensable ally in transforming my vision into this finished book.

SOURCES

Most of the information in this book has been compiled from papers published in geological and seismological journals and from reports prepared for various government agencies. The assessments of fault activity and estimates of future earthquakes all are taken from the most recent available technical sources. The principal ones are an early article by San Diego geophysicist and geologist Bob McEuen and Jerry Pinckney (1972); reports by the local engineering geology companies of Leighton and Associates (1983) and Woodward-Clyde Consultants (l986); and articles by Cal Tech geologist Steven Wesnousky (1986) and by SIO and SDSU geoscientists John Anderson, Duncan Agnew, and Tom Rockwell (1989). Complete references to these and other sources (which are indicated in the text by names or dates in parentheses) are in the Bibliography. Parts of the text were excerpted from a report (Kern, 1987) that I wrote for the San Diego County Office of Disaster Preparedness. What you read here, then, is not personal opinion, and it is not conjecture; it is the current concensus among professional geologists and seismologists who are authorities on earthquake hazard in San Diego County, and it is based on all the available geologic and historic evidence.

PHILIP KERN

The author received the doctorate in geology from the University of California at Los Angeles and is Professor and chairman of the Department of Geological Sciences at San Diego State University. He mapped faults for a decade as part of a more inclusive study of the Pleistocene geologic history of coastal San Diego County. His technical publications include articles on San Diego fossils, marine terraces, and Pleistocene deformation, and non-technical writings have appeared in Sierra, San Diego Home and Garden, Environment Southwest, and the San Diego Tribune.

CONTENTS

PREFACE

Earthquakes and Faults in San Diego was first published as a 32-page booklet in 1983; since then it has been reprinted twice with minor changes and additions. During these six years, though, there has been an explosion of knowledge about San Diego County's faults and earthquake history, and this knowledge has given geologists and seismologists a much better understanding of fault behavior and earthquake hazard throughout the County. That hazard turns out to be substantially greater than has generally been recognized.

Rather than try to incorporate bits of additional information into the original text, I have taken this opportunity to expand the scope of the book to the entire County; to add chapters on the nature of earthquake hazards, on geologic methods of estimating future earthquakes, and on earthquake preparedness; and to revise and update the entire book. Only the second and sixth chapters, plus much of the description of the Rose Canyon fault zone, have been retained with only small changes. The earthquake history is now complete through March, 1989.

The resulting 80-page book still is intended to serve as a non-technical guide to earthquakes and faults throughout San Diego County. I have attempted to present geologic concepts in such a way that they can be understood by someone with little or no prior knowledge in the science. The book includes a review of the major quakes that have shaken the County since 1800, maps of principal earthquake epicenters and of all major active faults, and summaries of probable future earthquake activity on these faults as it has been estimated by geologists and seismologists. The book also is intended to serve as a field guide to these faults, and there are descriptions and illustrations of easily accessible places where all of them can be seen.

Among the various hazards posed by the faults are the perils encountered in looking at them. Some of the exposures described here are in high, treacherous bluffs or even cliffs that readily shed boulders, people, and other objects. Some are adjacent to roads and streets that carry fast, heavy traffic; this is especially true at roadcuts on mountain and desert highways. When you go looking for faults, watch your step and keep an eye on vehicles, falling rocks, and other projectiles. Be especially careful at highway road cuts to stay out of traffic lanes and not to obstruct or distract drivers.

Take the family. Bring a lunch. The maps and pictures in this book will guide you through some of our faulted and sometimes quaking landscape. I hope you will gain from your excursions a   greater knowledge and understanding of earthquakes and faults in San Diego County and a better appreciation of our likely earthquake future and the measures one can take to prepare for it.

San Diego
April 1, 1989

EARTHQUAKES AND FAULTS IN SAN DIEGO COUNTY

I. THE EARTH QUAKES

February 23, 1892. The winter night is clear and cool on the coast, cold in mountains and desert. It is just before midnight and San Diego County sleeps.

Suddenly the ground heaves. Deep beneath the desert floor, near the southeastern corner of the County, two vast slabs of rock wrench abruptly past one another. The shock sends great waves rolling out through the planet's crust, tossing the landscape overhead as on spreading ripples in the sea. The ground is cracked for a distance of more than 12 miles along the Laguna Salada fault, and at one place a wall of rock rises twelve feet out of the desert floor.

At Carrizo Stage Station several adobe buildings are destroyed. A wall is shaken down at Campo, and others are cracked. Small objects are overturned at Julian and at Escondido, where goods are shaken off storeroom shelves. A stone kiln is cracked at Jamul, and in San Diego many buildings are cracked and ceilings unplastered, as frantic citizens are scattered into the streets.

This earthquake, which was felt in Needles, Visalia, and Santa Barbara, is estimated to have had a magnitude close to 7. It was the strongest quake recorded during historic times in or adjacent to San Diego County, and it produced stronger shaking there than has any other quake. San Diego, 70 miles away, was shaken nearly as strongly as it had been in 1862.

Because no quake so powerful has struck here again, many people think there is no earthquake hazard in San Diego. In fact, however, the 1892 temblor was not a freak event but was just one of three damaging quakes that struck during the nineteenth century. The first of those, on November 23, 1800, cracked adobe walls of the mission buildings both in San Diego and in San Juan Capistrano. To accomplish that feat, it must have had a magnitude of 6.5 or greater, and it probably was centered in the coastal zone between the two missions, perhaps in the vicinity of Oceanside.

Fault scarp on LAGUNA SALADA FAULT, Baja California, produced by movement in 1892 earthquake (from photo provided by Professor Tom Rockwell)

The most destructive quake recorded in the city of San Diego, though, was the one that struck the sleepy town of a few thousand people around midday on May 27, 1862. Pendulum clocks were stopped, and the Army Depot bell in Old Town was set ringing. Crockery smashed to the floor throughout the town, and walls were cracked in the Pico and Bandini Adobes, the Whaley House, and the stone lighthouse on Point Loma. Wooden buildings sagged, while door hinges and windows were shattered. Cracks gaped in the earth near the river and on the beach, and water surged up on the shore in both places. For several seconds the town continued to quiver. Then, after a few moments of quiet, there was a second shock; once more the shaking gradually subsided and finally the earth was still again, though aftershocks would continue into the following year.

From newspaper accounts, letters, and other written records the shaking and destruction in San Diego have been measured at VI to VII on the Modified Mercalli scale (described in chapter VI) of earthquake intensity. In San Diego shaking was a bit stronger than during the 1800 and 1892 quakes. A comparable earthquake today could result in considerable damage in poorly built or poorly designed structures. Unreinforced masonry walls, chimneys, and building cornices would be cracked and some would collapse. Much plaster and stucco would fall, and windows and furniture would be broken. There would also be prodigious breakage of dishes, glassware, knick-knacks, pictures, etc. Nearly everyone would be frightened, and some would find it difficult to stand. This would be a major earthquake with extensive damage.

The location of the 1862 quake remains uncertain. It may have been centered in the Rose Canyon fault zone directly under San Diego, in which case the damage it caused suggests a magnitude of about 6.0. Or it may have been offshore to the west or southwest of the city, perhaps in the Coronado Bank or San Diego Trough fault zone, and the magnitude would have been greater. The important point, though, is that, wherever its epicenter and whatever the magnitude, the damage it caused in San Diego was greater than in any earthquake experienced there during the present century. It is just as clear from this record of nineteenth century earthquakes that equally powerful future quakes can be expected throughout San Diego County.

In fact, the eastern and northern portions of the County have been shaken about as hard during the past two centuries by at least seven other earthquakes in addition to the three described above. Those quakes have received even less attention because these areas have relatively small population and few structures, and damage in those quakes has been minor.

Unfortunately the past century of relative quiescence in the populated areas has misled many into believing that there is no earthquake hazard here. Clearly this is not the case, and it is important that the misconception be corrected and that individuals, businesses, and governments alike prepare themselves for the inevitable earthquakes to come.

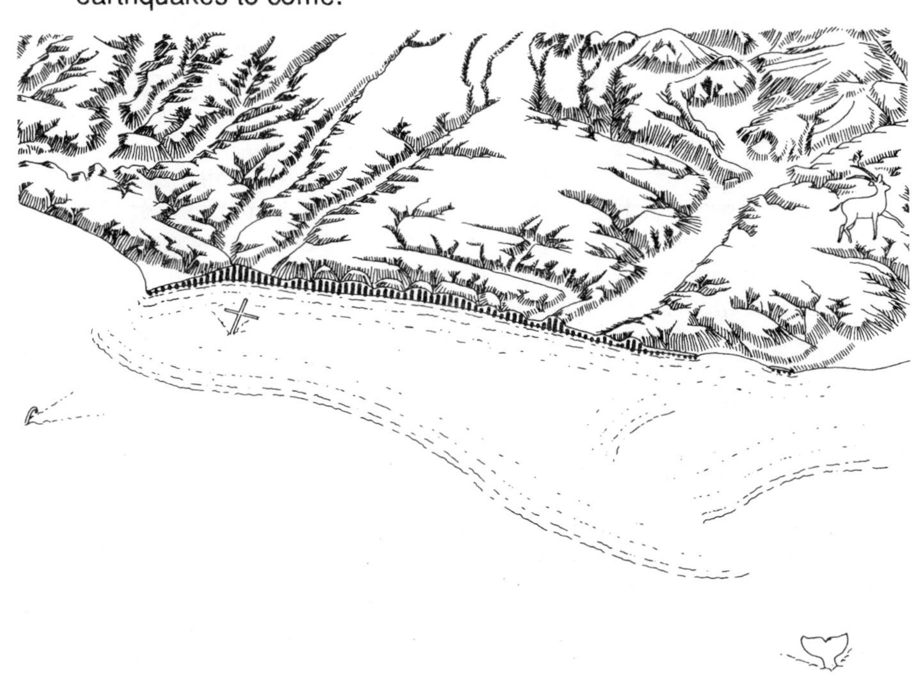

II. FAULTS AND EARTHQUAKES ARE CAUSED BY MOVEMENTS OF EARTH'S CRUSTAL PLATES PAST ONE ANOTHER

Earthquakes occur here because San Diego County occupies a broad belt of great instability in Earth's crusty skin. Deeper layers of the planet (in the mantle), because of extreme heat and pressure, are not solid but behave as a thick fluid, like cold honey, that flows about under the brittle crust. The crust is broken thereby into great plates, 40 to 50 miles thick, that float ponderously about on the slowly swirling currents beneath.

The boundaries between adjacent moving plates are fault zones, in which the faults that we see at the surface extend down through the crust. Friction prevents adjoining plates from moving past one another most of the time, but continual motion of their fluid underpinnings constantly increases the pressure that is exerted along the faults. When the resisting force eventually is overcome, the two plates lurch abruptly past one another, sending waves of motion -- an earthquake -- out across the planet.

Earth's crust under the Pacific Ocean is on one such plate, the crust of North America on another. The Pacific plate is gliding northward, at several inches each year, past North America. The boundary between these plates, however, is not a simple one. It consists rather of a broad belt that is broken into long, irregular slices by a series of more-or-less parallel faults. The accompanying map shows the chief fractures of this series, from the San Andreas fault on the east to the San Clemente fault on the west.

While many of the most devastating earthquakes in California have resulted from movements on the San Andreas fault, as in San Francisco in 1906 (magnitude 8.3) and at Fort Tejon in 1857 (magnitude 7.9), the entire zone is prone to convulsion. The San Jacinto fault, which is an Imperial Valley branch of the San Andreas, has spawned 13 quakes of magnitude 6 or higher in the past century. The Elsinore fault has had a somewhat less tumultuous recent history, with only one quake greater than magnitude 5 during the past 50 years. Similarly the Rose Canyon fault and those offshore as far as San Clemente Island also remain active. San Clemente Island had a quake of magnitude 5.9 in 1951, and in 1986 there was another of magnitude 5.3 on the Coronado Bank fault off Solana Beach. San Diego County thus lies within a broad belt of shattered crust that is subject throughout to violent quaking as its western margin jerks northward past the eastern one.

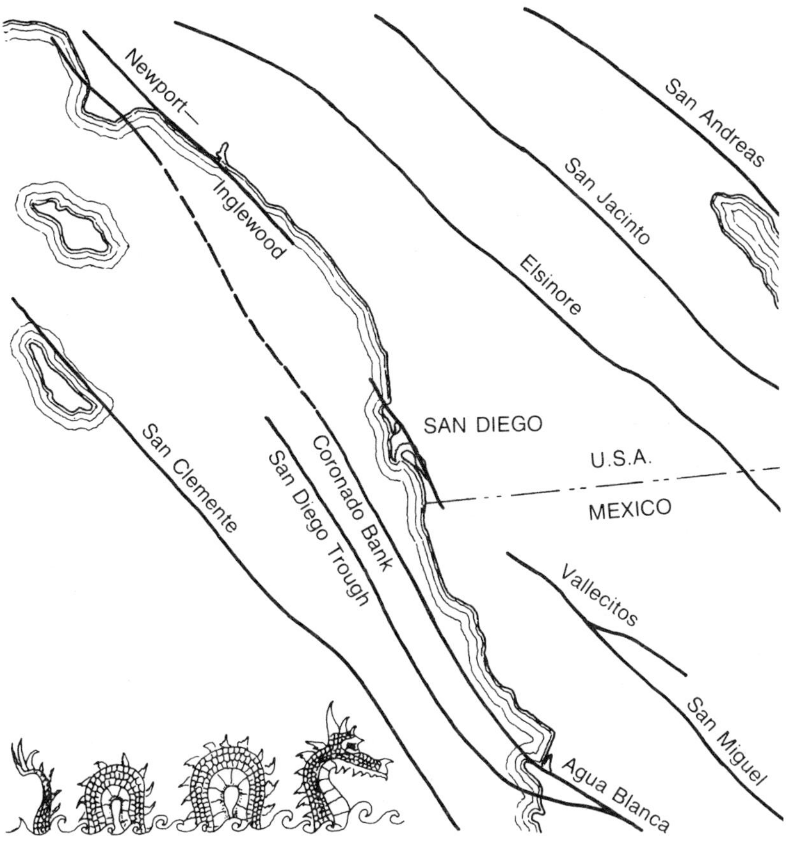

Active faults of San Diego County

III. EARTHQUAKE MAGNITUDE IS RELATED TO FAULT LENGTH, LENGTH OF FAULT RUPTURE, MAXIMUM AMOUNT OF FAULT MOVEMENT, RATE OF MOVEMENT, AND TIME INTERVAL BETWEEN QUAKES

The point at which the initial rupture occurs on a fault, and from which the earthquake waves radiate outward, is called the focus, or hypocenter. The epicenter is the point on the earth's surface directly above the focus. If the distance the rocks move past one another (the amount of slip) is large at the focus, say several to many feet, then much energy is released and the earthquake is of large magnitude. The fault plane also will be ruptured (will slip) over a large part of its length and depth. It may rupture at the ground surface for distances up to hundreds of miles. If on the other hand the earthquake is of small magnitude or its focus is deep, the rupture may not reach the surface at all.

The amount of slip varies along the fault plane, with a maximum amount at the focus decreasing in all directions to zero at the edges of the zone of rupture. In the 1906 San Francisco earthquake (magnitude 8.3), for example, the San Andreas fault ruptured at the surface for 270 miles, from near Cape Mendocino to San Juan Bautista. The largest amount of horizontal slip at the surface was measured in Marin County, where a road and a fence were offset by 20 feet, while at San Juan Bautista slip ranged from a few inches to zero. In the more powerful 1964 Alaska earthquake (magnitude 8.4) the fault ruptured at the surface for 500 miles.

Generally speaking, then, higher magnitude results from longer surface rupture and greater maximum amount of slip. Though it varies considerably with the kinds of rocks, depth of earthquake focus, and other factors, the relationship between earthquake magnitude, length of surface rupture, and maximum amount of slip is approximately as follows:

MAGNITUDE	RUPTURE LENGTH	MAXIMUM SLIP
5.5	2 miles	1 inch
6.0	6	3
6.5	12	1 foot
7.0	30	3
7.5	75	6
8.0	210	33
8.5	600	100

Thus longer faults are capable of producing more powerful earthquakes, and one way of estimating the earthquake hazard of a fault is to determine the maximum magnitude that is likely if it ruptures along its entire length.

The amount of slip that occurs, and resulting earthquake magnitude, also are controlled by a fault's average long-term rate of slip, generally measured by geologists in millimeters per year, and its average earthquake recurrence interval -- the time interval between earthquakes. The long-term average slip rate is controlled by the velocity at which the plates are being pulled along by underlying currents, while recurrence interval is influenced by the kinds of rocks, the direction of fault motions, the geometry of the fault plane, and other factors that determine how much stress can build up before rupture occurs. If slip rate is high (more than a few millimeters per year) and recurrence interval is long (hundreds of years or more), a fault will be characterized by infrequent but strong earthquakes. If slip rate is low (less than a few millimeters a year) and recurrence interval is short (tens of years), the fault will be characterized by frequent smaller quakes.

Faults tend to have a characteristic behavior in these regards, and, if their earthquake history is known, then the long-term average earthquake pattern is more-or-less predictable. Many faults seem to have "characteristic" earthquakes all of about the same magnitude, and with a fairly consistent average recurrence interval.

Long faults, though, are unlikely to rupture throughout in a single earthquake. Rather, they tend to be segmented, with the segments separated from one another by bends, branches, or discontinuous steps to the left or right. It is thought that these discontinuities are surface expressions of deep-seated obstacles to rupture. Earthquakes typically result from rupture of just one or, in some instances, a few segments; thus it is often fault segments whose characteristic earthquake magnitude and recurrence interval are estimated.

Some faults or fault segments may produce earthquakes of magnitude greater than their length would suggest because they become "locked". If a fault bends or steps in such a way that movement is impeded, stress may accumulate for a much longer time than it would on a straight fault. Unusually long recurrence intervals then would be punctuated by unexpectedly powerful earthquakes.

Some faults also appear to have extraordinarily long recurrence intervals -- as high as thousands of years and perhaps even tens of thousands of years. It is thought that rocks on opposite sides of faults with exceptionally low slip rates may become annealed, or fused together, with the result that stress builds up to a much higher degree than it normally would. Earthquake magnitudes far out of proportion to fault length can result at just as unexpectedly long intervals. Where such conditions exist, it may be extremely difficult to know when faults have truly become inactive.

Some faults or fault segments "creep", or move more or less continuously; some of these creeping faults also have earthquakes, while others have no quakes at all. A 60-mile-long segment of the San Andreas fault from Hollister to west of Coalinga has been creeping at about one inch a year during historic time.

IV. MAGNITUDES OF FUTURE EARTHQUAKES ARE ESTIMATED FROM THE HISTORIC RECORD AND FROM GEOLOGIC EVIDENCE FOR THE PREHISTORIC RECORD

Earthquake hazard evaluation requires estimates both of probable magnitude of future earthquakes and of average frequency, or recurrence interval, at which they are likely to occur. This information can be obtained from the historic record, if that record is long enough, or it can be determined from geologic evidence of the prehistoric record. Estimation of approximate locations, magnitudes, and times of occurrence of future earthquakes allows for anticipation of shaking and other damage in any particular place depending on its distance from the earthquake and the local geologic conditions.

The most direct evidence of earthquake hazard is that of the historic record -- written accounts in old newspapers, letters, diaries, journals, mission records, and so on. Residents of San Francisco can expect eventually to experience a repetition of the 1906 earthquake, as those of Anchorage can anticipate another like that of 1964. The sparse accounts we have of the San Diego County earthquakes of 1800, 1862, and 1892 warn us to be prepared for comparable quakes in the future.

If the historic record is long enough to reveal the long-term earthquake behavior of a fault, one can estimate probable magnitude and approximate recurrence interval of future quakes, as faults seem to have "characteristic" earthquakes of about the same maximum magnitude and at more-or-less regular intervals. The principal uncertainty would be in the timing. Some faults have short average recurrence intervals of a few tens of years, while others have quite long intervals, up to at least several hundred years and perhaps in some cases even thousands or tens of thousands of years. There is no way of determining such long intervals from the historic record, as that record is too short. Faults with such long recurrence intervals would be unlikely to have had an earthquake in historic time, so even their characteristic magnitude would not be known directly.

Recurrence intervals are rather uniform on some faults, while on others they may vary widely about the average. One of the more regular southern California faults, for example, is the San Jacinto, which passes through the northeastern corner of San Diego County. It has had 13 earthquakes of magnitude 5.8 to 6.8 since 1899, an average of one every 8 years and with no interval greater than 16 years. Because most faults have much longer recurrence intervals and the historic record in southern California goes back only to around 1800, that record provides little or no information on earthquake recurrence intervals on some of our major faults.

Thus there are limitations to the historic record, but it does provide some information on earthquake recurrence interval, approximate magnitude, fault rupture length, and long-term rates of slip. Where data are well documented, these facts serve as a basis for estimation of a fault's probable future earthquake hazard.

Geologic evidence for the prehistoric earthquake record comes from features such as fault scarps (cliffs formed by movement of one side of the fault above the other side), as in the illustration of the Laguna Salada fault in chapter I; offset streams or rock formations; and displaced beds seen either in natural outcrops or in trenches dug specifically for the purpose of studying faults. Stream channels, for example, are offset in crossing many of southern California's active faults. In some cases one channel approaches a fault and several leave it on the other side, the separate offsets recording movements in successive earthquakes. Amounts of slip indicated by such offsets can be used to estimate approximate magnitudes of future quakes. If the dates of the past earthquakes can be determined, say by carbon-dating of deposits into which the stream channels are cut, then one can determine earthquake recurrence interval and long-term rate of slip

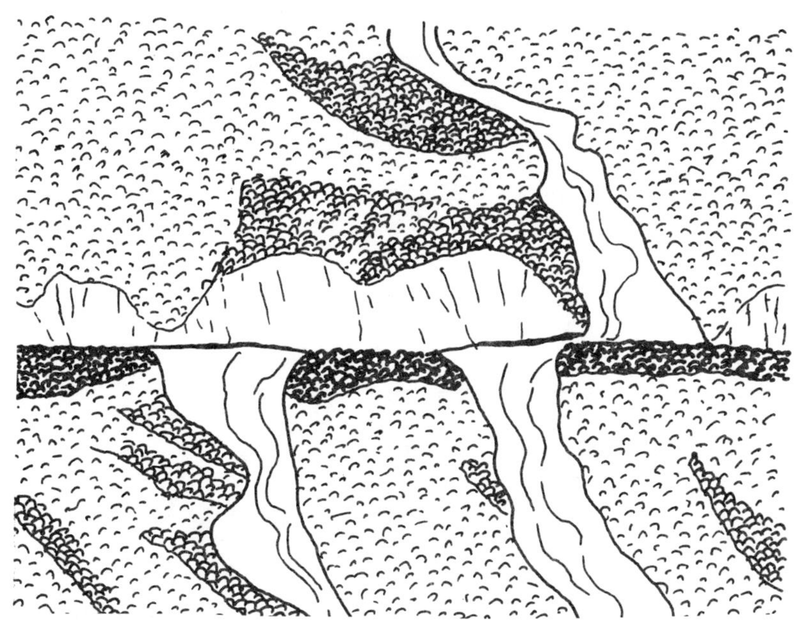

Oblique air view of stream valley offset twice by earthquakes on SAN JACINTO FAULT just east of San Diego County (from photo provided by Professor Tom Rockwell; width of view one-quarter mile)

on the fault. Such slip rates also can be calculated if one knows the age of a rock formation that has been displaced a known distance along a fault.

Another method of obtaining information on prehistoric earthquakes is by digging trenches across faults that cut through deposits of sediments that have been accumulating during recent times, as in ponds and swamps. In such trenches one commonly finds that successively lower, thus older, layers of sand and mud have increasingly greater offsets because they have been displaced by more earthquakes. Here it can be even simpler to determine the dates of those earthquakes by carbon-dating of wood, shell, or other organic materials that were buried in the displaced sediments. Magnitudes of the earthquakes are suggested by the amount of slip, and here too it may be

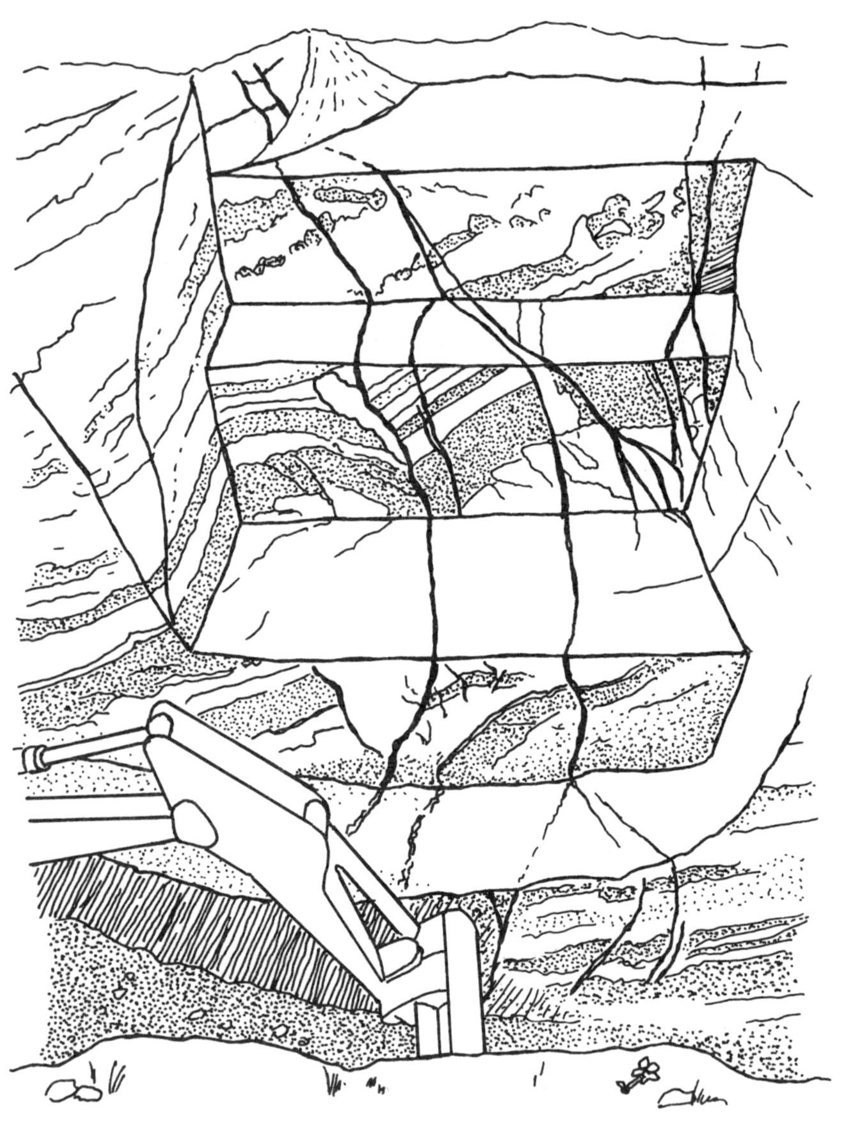

ELSINORE FAULT in trench at Glen Ivy Marsh north of Lake Elsinore (from photo provided by Professor Tom Rockwell)

possible to reconstruct earthquake history -- dates, amounts of slip, magnitudes, long-term average slip rates, and average recurrence intervals.

Among a variety of odd sources of information that geologists have found are old trees that are growing on a fault and that were injured or killed by earthquake movements. The date of the disturbance can be obtained from the tree-ring record by comparing it with rings in nearby trees not on the fault. By using these and other kinds of evidence for the longer prehistoric record, it is possible to increase the base of information from which future earthquake risk can be estimated.

Where there is no information on age or magnitude of historic or prehistoric earthquakes, it is more difficult to evaluate earthquake hazard. If there is topographic evidence of recent activity, say in fresh fault scarps, offset streams, or other fault features, one can infer that there is some future earthquake hazard. The relationship between earthquake magnitude and fault length or fault segment length then allows one to make at least a general estimate of likely magnitude.

In any case there is always some uncertainty in estimation of earthquake hazard. Recurrence intervals can vary widely, and faults can alter their behavior. Active faults can become inactive, their previous movements taken up later on other fault zones, and new faults can appear. Important fault features, or entire faults themselves, may not be clearly exposed and thus may remain completely unknown until an earthquake occurs. All the estimates in the following chapters must be regarded as exactly that -- estimates of what appear to geologists and seismologists, from existing knowledge, to be the most likely future behavior of the fault in question.

V. EARTHQUAKE DAMAGE VARIES WITH MAGNITUDE, DISTANCE FROM EPICENTER, GEOLOGIC CONDITIONS, AND BUILDING CONSTRUCTION; AND IT INCLUDES THE EFFECTS OF GROUND SHAKING AND RUPTURE, LANDSLIDES, TSUNAMIS, AND FIRES

The principal factors that determine the shaking damage caused in a particular place by an earthquake, in addition to magnitude, include the distance from the epicenter, the duration of shaking, the local geologic conditions, and the type and quality of building construction.

Just as the ripples spreading from a stone thrown in a pond grow smaller as they travel farther away, earthquake waves also decrease in size and energy as they spread farther from the focus. The farther one is from the epicenter, the smaller are the waves and the less is the shaking and damage. The 1906 San Francisco earthquake of magnitude 8.3 was felt from Oregon to Los Angeles, a distance of 700 miles, and 350 miles to the east in Winnemucca. Severe destruction, though, was restricted to the area from Eureka to Santa Cruz; damage was great but not as extensive between Monterey and King City; in the Coalinga area damage was slight; and there was no damage at all south of San Luis Obispo. The isoseismal map shows how shaking and damage diminished in concentric rings outward from the epicenter of the 1892 quake on the Laguna Salada fault. The numbers on the map are Mercalli intensities.

Damage also is related to duration of shaking, which can vary from a few seconds to as much as several minutes. Much more damage is done if shaking is prolonged for more than a few seconds, even though the intensity of shaking is the same. In both the San Francisco (1906) and Alaska (1964) earthquakes, intense shaking continued for at least half a minute.

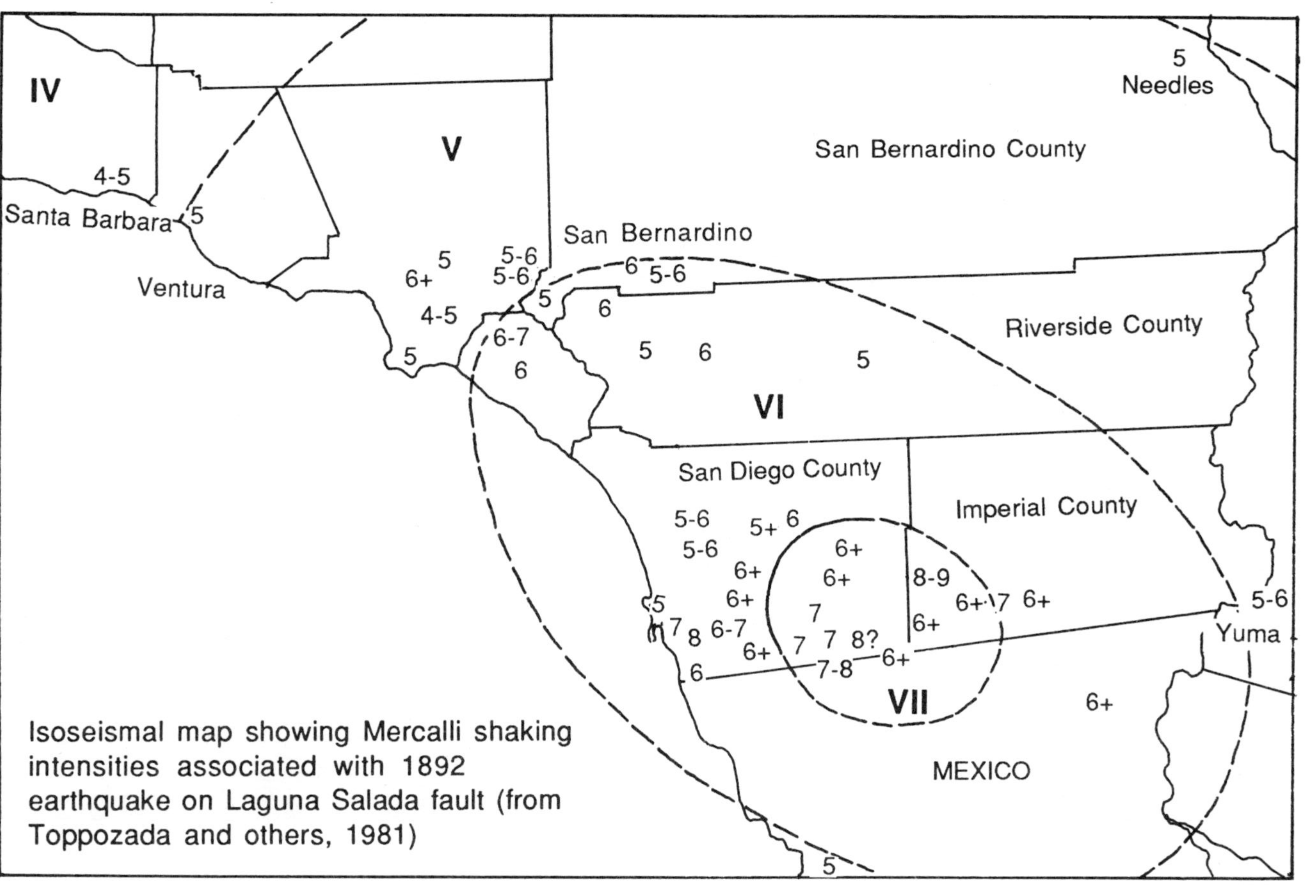

Isoseismal map showing Mercalli shaking intensities associated with 1892 earthquake on Laguna Salada fault (from Toppozada and others, 1981)

The amount of shaking and consequent damage in a given place also is strongly influenced by the nature of the ground there. In the 1985 Mexico City earthquake some buildings built on loose, water-filled sands were totally destroyed, while others built on bedrock literally across the street were undisturbed. The epicenter, incidentally, was on the coast some 200 miles away.

The greatest shaking, and consequent damage, occurs on loose, unconsolidated sediments, especially if they are saturated with water. Water-saturated sediments are subject to liquefaction, a process in which shaking turns a coherent mass of wet sediment into jelly, causing the ground to flow and crack and buildings to topple. Unconsolidated sediments on hillsides also are prone to landslides, again especially if they are water-saturated.

Soft or poorly consolidated rocks shake less than loose sediment, and hard, solid rocks are even more stable. Granite, basalt, limestone, and similar rigid bedrock formations are the least subject to earthquake shaking.

Shaking damage to structures also varies dramatically according to type and quality of construction. Rigid structures of masonry, stone, or adobe are more susceptible to shaking damage than are more flexible wooden frame buildings. Unreinforced brick and masonry structures, including walls, chimneys, and building parapets, are especially vulnerable. So are both older wooden frame buildings that are not reinforced or bolted securely to reinforced foundations and newer frame buildings that have large windows or poorly reinforced garages underneath.

While most earthquake destruction results from shaking, surface rupture of faults also can cause considerable damage. If buildings, roads, bridges, aqueducts, pipelines, or other structures are built across faults, ground rupture of those faults can be devastating, although relatively few structures usually are involved.

Tsunamis, or seismic sea waves, can be triggered by strong earthquakes beneath the sea, and they pose potential danger to coastal areas. Coastal San Diego County appears, however, not to be highly vulnerable to this hazard, apparently because of the configuration of the coastline relative to the likely sources of such waves. There have been no damaging tsunamis here during historic time.

Among a variety of other hazards associated with earthquakes are ground subsidence, oscillation of water in lakes or rivers (seiches), changes in water table, floods from dam failure, and fires. Generally these and other effects are minor, but they can be disastrous, as shown by the 1906 San Francisco fire.

In addition to the extensive damage that strong earthquakes might cause in homes, businesses, and other private buildings, there may be a much greater threat to the community in the damage that can be done to public structures. These include buildings -- such as hospitals, schools, and police and fire facilities -- and

Earthquake damage; intensity VII (from photo provided by Professor Rick Miller)

transportation and utility lifelines -- airports, highways, railroads, harbors, telephone lines, electric transmission lines, gas pipelines, water supply lines, sewage lines, and dams. The conclusions of a recent study (Reichle and others, 1989) of such effects from a projected earthquake of magnitude 6.8 between San Diego and Tijuana give an idea of what might be expected there, and similar effects could be anticipated throughout the County from earthquakes in other areas.

Liquefaction of filled ground and natural waterfront deposits probably would close I-5, the I-8 to 163 interchange, and the Coronado Bridge; similar closures could occur in lagoon and river-bottom areas along the entire coast. The runways at Lindberg Field could be closed for as long as 72 hours because of liquefaction damage. Rail service also could be interrupted by track damage because of liquefaction in many of the coastal valleys. Highways and airports are less likely to be damaged on the firmer ground of mesas and inland areas.

Telephone communications also could be disrupted because of shaking damage to lines and switching equipment as well as overloading by post-earthquake calls. Power outages probably would be limited in extent in most areas, but damage from liquefaction around San Diego and Mission Bay, and in the smaller developed waterfront areas up the coast, might put power out locally for 24 to 72 hours. Major water supply aqueducts should not be seriously affected by a coastal quake because they are all on firm inland ground, but strong earthquakes on the Elsinore or San Jacinto zones could cut those lines for several days. Local water lines within towns and cities could of course suffer extensive damage from any strong nearby quake. Natural gas pipelines enter San Diego from the north along the coast, so they would be subject to liquefaction and landslide damage from any strong quake and could be out of service for as long as 72 hours.

Thus individuals and business and community leaders should make preparations for the possibility of going without water, electricity, natural gas, and telephone services for several days in the event of a major earthquake. They should also be aware that firefighting efforts could be severely hampered during the first 72 hours because of broken water lines and blocked streets, and medical and other emergency aid could be similarly hampered as well as overwhelmed by requests for assistance.

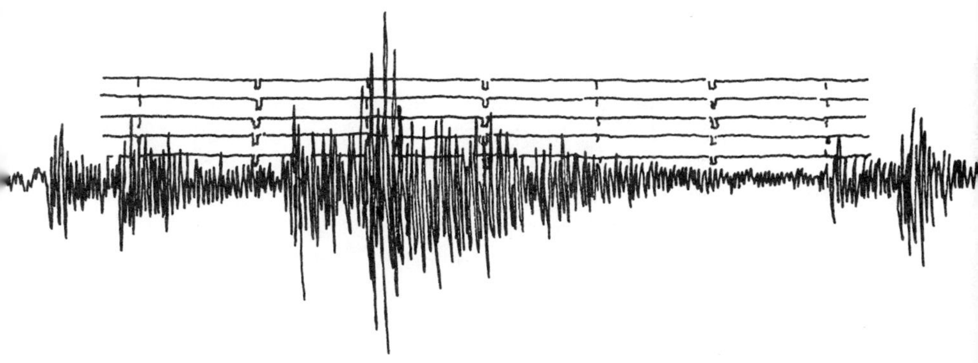

SDSU seismogram for the 1987 Whittier earthquake. Three successive shocks are shown, each with an initial primary wave followed by a larger secondary wave. The 20-second delay between their arrivals (there are 30-second marks on the other lines) is the difference in their travel times from Whittier to San Diego

VI. EARTHQUAKES ARE MEASURED IN MAGNITUDE AND INTENSITY

Seismographs are instruments that record precisely the arrival times of the several different kinds of seismic waves that travel at different velocities from the focus of an earthquake. The lengths of the time intervals between those arrivals indicate the distance to the focus, and distances measured from three different seismograph stations can be used to locate the point of origin by triangulation.

Seismographs also measure the distance that the earth's surface is moved by the passage of seismic waves. The seismograph magnifies the earth movement, and a pen traces out a seismogram like the one on the cover of this book. The greater the magnitude of an earthquake, the larger the waves it dispatches, and the wider the swing of the seismogram line. Operating seismographs can be seen at the Natural History Museum in Balboa Park, in the gift shop museum at Scripps Institution of Oceanography, and in the Department of Geological Sciences at San Diego State University.

There are two kinds of scales for measuring and comparing earthquakes -- one of magnitude and one of intensity. The magnitude scale is the one that is based on seismograph measurement of ground vibration and which thus indirectly measures the energy released at the focus of the quake. Best-known and most widely used is the Richter scale. It is based on a logarithmic comparison of the seismograph-measured heights (amplitudes) of earthquake waves. Each step on the scale thus marks a tenfold increase in height of the waves that pass through the ground. Even though the waves represent only a few per cent of the total energy released by an earthquake, a quake of magnitude 8 radiates not just ten, but more than thirty times the energy of one of magnitude 7, and it has a thousand times as much seismic energy as one of magnitude 6.

It is a common misconception that the Richter scale extends from one to ten. In fact it has no lower and upper limits. Quakes of very little energy actually have minus magnitude numbers. The smallest quakes detectable by seismographs have magnitudes near -3, while those of magnitude 2.5 to 3.0, of which there are approximately 100,000 annually, are the weakest that can be felt by people in the immediate area. The energy released by the first nuclear bomb explosion was about that of an earthquake of magnitude 5, but severe destruction from earthquakes begins only near magnitude 6. A magnitude 7 quake can be detected around the entire planet, and about 15 to 20 such major shocks occur in an average year.

It has been found recently that the Richter scale is not accurate for measuring earthquakes stronger than about magnitude 8, and a modified scale has been developed to measure these more powerful quakes. The 1906 San Francisco earthquake has been downgraded to magnitude 7.9 on this scale, while the 1964 Alaska quake has been upgraded to 9.2, the second strongest known earthquake. The strongest of all was a 1960 Chilean quake, which measured 9.5 on the new scale.

Magnitude numbers commonly are indicated by a capital M, as in M5 (magnitude five).

The Mercalli scale of earthquake intensity is used to indicate an earthquake's apparent severity at a particular location, and it is based on the observed effects of the shaking produced by the quake. Intensity is determined from damage surveys and from descriptions of shaking effects by people in the area. Intensity of past earthquakes can be estimated from written accounts in newspapers, letters, diaries, and other such sources.

While magnitude is an indirect measure of energy produced at the focus and has only one value for a given earthquake, intensity decreases with increasing distance from the epicenter and with variations in local geologic conditions. Intensity numbers usually are indicated by two capital M's, as in MM VII (Modified Mercalli intensity of seven). Following is an abbreviated form of the modified Mercalli scale of felt earthquake intensity. More complete descriptions can be found in Bruce Bolt's (1978) book and in many other sources.

I -- Felt only by a very few people under exceptionally favorable circumstances.

II -- Felt only by a few people at rest, especially on upper floors. Suspended objects may swing.

III -- Felt distinctly indoors, especially on upper floors. Standing vehicles may rock slightly. Vibration like that of a passing truck.

IV -- Felt indoors by many but outdoors by few. Some people awakened at night. Standing vehicles rock noticeably. Dishes, windows, and doors disturbed. Walls make cracking sound. Sensation like that of a truck striking the building.

V -- Felt by nearly everyone. Many awakened. Trees, buildings, and other tall structures sway. Small objects overturned. Some dishes and windows broken. Some cracked plaster. Pendulum clocks may stop.

VI -- Felt by all. Many frightened and run outdoors. Some heavy furniture moved. Some fallen plaster or damaged chimneys. Damage slight.

VII -- Everyone runs outdoors. Damage negligible in well-designed and well-built structures; slight to moderate in ordinary well-built structures; considerable in badly designed or poorly built structures. Some chimneys broken. Felt by people in moving vehicles.

VIII -- Damage slight in specially designed structures;
considerable in ordinary substantial buildings; great in
poorly built structures. Panel walls thrown out of frame
structures. Chimneys, factory stacks, columns,
monuments, and walls toppled. Heavy furniture
overturned. Some sand and mud ejected. Changes in well
water. Disturbs people in moving vehicles.

IX -- Damage considerable in specially designed
structures. Well-designed frame structures thrown out of
plumb. Buildings shifted off foundations. Ground cracked
conspicuously. Underground pipes broken.

X -- Some well-built wooden structures and most masonry
and frame structures destroyed. Ground badly cracked.
Rails bent. Many landslides from river banks and slopes.
Shifted sand and mud. Water splashed over banks.

XI -- Few if any masonry structures remain standing.
Bridges destroyed. Broad fissures in ground. Underground
pipelines completely out of service. Earth slumps and
landslips in soft ground. Rails bent greatly.

XII -- Damage total. Waves seen on ground surfaces.
Lines of sight and level distorted. Objects thrown upwards
into the air.

The following table shows approximate relationships
between earthquake magnitude and intensity at the
epicenter. Intensity decreases with increasing distance
from the epicenter.

M	MM
3	III
4	IV-V
5	VI-VII
6	VII-VIII
7	IX-X
8 +	XI-XII

VII. MANY STRONG EARTHQUAKES HAVE STRUCK SAN DIEGO COUNTY IN THE PAST TWO CENTURIES

The following table lists all the earthquakes that have produced strong shaking or damage in San Diego County since 1800. No records of strong earthquakes are known from before that time. The list has been compiled principally from the two books by Toppozada and others (1981, 1982). The table gives dates, magnitudes (estimated for dates before 1934, when seismographs were first used here), epicenter locations, and intensities and damage from various places in the County. Question marks indicate that magnitude or epicenter location are not known or that information given is uncertain. Notice that many of the earthquakes listed here, especially those of large magnitude, had epicenters outside San Diego County. The fault map in chapter VIII shows epicenters of those quakes of magnitude 4.5 or greater that have struck in or near San Diego County or in the offshore area.

STRONG EARTHQUAKES IN SAN DIEGO COUNTY

1800   Nov. 22. M6.5? Coastal San Diego County. MM VII at San Diego and San Juan Capistrano; Adobe walls cracked.

1803   May 25. ? ? MM VI at San Diego; mission church damaged slightly. MM VI-VII inland.

1812   Dec. 8. ? San Andreas fault. Strong shaking but no damage at San Diego. San Juan Capistrano Mission church destroyed; 40 killed.

1852   Apr. 12. ? ? MM VII in San Diego. One home destroyed but no other damage.

1852   Nov. 29. ? Mexicali Valley. MM V in San Diego.

1856   Sep. 20. ? San Diego County. MM VII at Santa Ysabel; walls cracked and ceilings fell. MM VI in San Diego; windows rattled and small objects upset.

1857   Jan. 9. M7.9. Fort Tejon. MM V in San Diego; minor damage.

1859   Mar. 25. ? ? MM V at Ballena; cracked ground.

1862 May 27. M5.9? Coastal San Diego County. MM VII
       in San Diego. Buildings and wet ground cracked;
       door hinges, windows, and crockery broken;
       pendulum clocks stopped; bell rang at Army Depot.
       MM V-VI at Temecula and Aguanga; plates rattled
       and objects upset. Roof shingles cracked at Mesa
       Grande. Strong at Vallecitos and felt at Anaheim.
       Many aftershocks into 1863, especially May 29,
       June 13, and Oct. 21. Strongest quake in San
       Diego's recorded history.
1875 Nov. 15. ? Imperial Valley. Destroyed buildings
       in Mexicali. MM VI in Campo; furniture upset and
       dishes knocked off shelves. MM IV in San Diego;
       clocks stopped.
1885 Sep. 13. ? ? MM V in San Diego.
1886 Oct. 8. ? ? MM V in San Diego.
1890 Feb. 5. ? ? MM V in San Diego; dishes rattled.
1890 Feb. 9. M6? San Jacinto, Riverside County? MM
       V throughout San Diego County.
1891 Jul. 30. M7. Colorado River delta. MMVI in Yuma.
       MM V in San Diego; clocks stopped, china rattled,
       and furniture moved.
1892 Feb. 23. M6.7-7.3. Northern Baja California.
       MM VIII-IX in east County; adobe buildings
       destroyed at Carrizo Stage Station. MM VII at
       Jacumba, Borrego Springs, Campo (collapsed and
       cracked walls and goods off shelves), and San Diego
       (many buildings cracked and chimneys and plaster
       fallen). MM VI at Julian (light objects upset) and
       Escondido (objects overturned and goods off
       shelves). Felt in Visalia and Santa Barbara.
1894 Oct. 23. M5.5-5.7. Mountains east of San Diego.
       MM VII in mountains. MM V in San Diego; people
       fled from creaking and cracked buildings, and
       plaster fell. strong at Escondido.
1899 Jul. 22. M6.5. Cajon Pass. MM V throughout San
       Diego County.
1899 Dec. 25. M6.6-7.0. San Jacinto, Riverside
       County. MM VII at Agua Tibia Ranch and Warner
       Springs. MM VI at Ramona, Escondido, and
       Jacumba. MM V-VI in San Diego; clocks stopped.

1903 Jan. 23. M7. Colorado River delta. MM V in San
      Diego County.
1906 Apr. 19. M6. Imperial Valley. MM V in San Diego,
      Ramona, and Julian.
1915 Nov. 20. M7.l. Near Cerro Prieto, Baja California.
      MM V-VI in San Diego.
1918 Apr. 21. M6.8. Near San Jacinto. MM VII at
      Warner Springs and Valley Center. MM VI at
      Fallbrook, Mesa Grande, Julian, Escondido,
      Oceanside, and San Diego; clocks stopped and small
      objects upset
1919 Dec. 31. ? ? MM VI at Warner Springs (adobe
      walls cracked) and Mesa Grande.
1920 Oct. 5. ? ? Trees shaken and everyone ran
      outdoors at Warner Springs.
1927 Aug. 14. ? Near Barrett Reservoir. MM V+ at
      Alpine and Jacumba.
1929 Dec. 2. ? Near Ensenada. MM V+ at San Diego.
1934 Dec. 31. M7.1. South of Calexico. MM V-VI in San
      Diego; buildings cracked, plaster fell, and windows
      broken. MM IV in Solana Beach, Ramona, and
      Escondido.
1937 Mar. 25. M6.0. County line north of Borrego
      Springs.
1939 May l. M5. 60 miles southwest of San Diego. MM
      V in San Diego.
1940 May 18. M6.7. Imperial fault, Imperial County.
      MM V in Borrego Springs. MM IV in San Diego.
1942 Oct. 21. M6.5. Near Borrego Valley. MM VII in
      Carrizo Gorge. MM VI in Warner Springs, Santa
      Ysabel, Jacumba, and Campo; plaster cracked and
      glass broken. MM V at Aguanga, Escondido, Mesa
      Grande, Mount Laguna, and Oceanside.
1949 Nov. 4. M5.7. Baja California. MM VI in San Diego
      and Campo; large buildings and trees swayed, some
      buildings cracked, and people ran outdoors.
1951 Dec. 25. M5.9. San Clemente Island. MM VI in San
      Diego and Del Mar; goods off shelves and plaster
      cracked. MM V at Mount Laguna, Pala, and Barrett
      Dam. MM IV at Campo, Escondido, Jamul, Julian,
      Leucadia, Oceanside, Santa Ysabel.

1954 Mar. 19. M6.2. Santa Rosa Mountains. MM VI in
Borrego Springs, Warner Springs, Jamul, and San
Diego; plaster cracked and glass broken. MM V in
Aguanga, Campo, Julian, Pala, Ramona, Escondido,
Del Mar, and Oceanside.

1956 Feb 9. M6.8. Baja California. MM VI in Campo
and San Diego; plaster cracked. MM V in Cardiff,
Carlsbad, Del Mar, Poway, Escondido, and Pala.

1958 Jan. 14. M3.7. Chula Vista. MM V in San Diego.

1964 Jun. 21. M3.7. San Diego Bay. MM VI in San
Diego; slight damage.

1964 Jun. 22. M3.6. San Diego Bay. MM VI; a
typewriter bounced off a table, water splashed in a
tub, and fire alarms were activated.

1964 Dec. 22. M5.6. Offshore Ensenada; MM VI in
Imperial Beach and San Diego; slight damage.

1968 Apr. 9. M6.5. Ocotillo Wells. MM VII at Borrego
Mountain and Ocotillo Wells. MM VI in Borrego
Springs, Campo, Julian, Alpine, Ramona,
Escondido, Del Mar, and Encinitas. MM V in La
Jolla.

1979 Oct. 15. M6.6. Imperial fault, Imperial County.
MM VI in San Diego.

1983 Jun. 29. M4.6. 12 miles west of Tijuana. MM V in
San Diego.

1984 Jul. 1. M4.3. West of Solana Beach.

1985 Jun. 17. M4.2. San Diego Bay. MM VI in San
Diego.

1986 Jul. 13. M5.3. 25 miles west of Solana Beach.
MM VI from San Diego to Oceanside; broken
windows and waterlines; cracked walls, patios, and
sidewalks; fallen tiles; $400,000 damage.

1986 Jul. 31. M4.1. 25 miles west of Solana Beach.

1986 Oct. 28. M4.7. Southeast San Diego. MM VI in San
Diego.

1987 Nov. 23 and 24. M6.2 and 6.6. Superstition Hills,
Imperial County. MM VI in Borrego Springs,
Warner Springs, Jacumba and Jamul. MM V in
Campo, Julian, Ramona, and Escondido.

VIII. SAN DIEGO COUNTY'S MANY ACTIVE FAULTS CAN BE EASILY SEEN, AND MAGNITUDES OF THEIR FUTURE EARTHQUAKES CAN BE ESTIMATED

INTRODUCTION

The map shows the major active faults and fault zones and epicenters of strong historic earthquakes in San Diego County and offshore as far as San Clemente Island. Though locations are approximate, especially for quakes that occurred before 1934, notice that the epicenters are concentrated in the fault zones.

Three kinds of information are presented in the text for each of these faults. There is a description of each zone, together with information on and illustrations of locations where it can be easily seen. This part of the text can serve as a field guide to San Diego County's active faults. There is also an account of each fault's earthquake history and of what is known about slip rates, recurrence intervals, and other seismic characteristics. Finally there is a discussion of future earthquake potential, as best it can be estimated from historic records and from geologic evidence.

These estimates of future earthquake potential are taken from the most recent technical publications on the subject. The principal sources, with full references given in the Bibliography, are McEuen and Pinckney (1972), Leighton and Associates (1983), Wesnousky (1986), Woodward-Clyde Consultants (1986), and Anderson, Agnew, and Rockwell (1989). A more complete and more technical discussion can be found in my report to the County Office of Disaster Preparedness (Kern, 1987). The estimates of future earthquake magnitude and shaking intensity are summarized in chapter IX.

OFFSHORE FAULTS

Fault descriptions. The continental borderland west of San Diego is crossed by a complex system of faults that are continuous with onshore faults in northern Baja California and in coastalmost southern California. Both offshore and onshore parts of this system pose potential earthquake hazards to San Diego County. The major faults offshore are the Coronado Bank, San Diego Trough, and San Clemente. The Coronado Bank fault is continuous to the north with the Palos Verdes Hills fault and to the south with the Agua Blanca fault south of Ensenada. The latter also branches northward to connect with the San Diego Trough fault. The San Clemente fault comes ashore farther south as the San Isidro fault zone.

The Agua Blanca fault zone comes ashore at Punta Banda, on the south edge of Bahia de Todos Santos at Ensenada. The Point itself extends seaward as a fault block within the zone and is bracketed by the Agua Blanca fault itself on the north and the Maximinos fault on the south. Though several north-south faults within the zone are exposed on the north shore of Punta Banda, the Agua Blanca fault lies just offshore and passes under the south end of the estuary. Inland the fault zone follows Punta Banda Ridge into Valle Santo Tomas and Valle Agua Blanca, but nowhere are there clear exposures where the fault can be easily seen.

Earthquake history. Recent activity is indicated on all three of the offshore fault zones by ocean-floor scarps, offset submarine canyons, and offsets of uppermost ocean-floor sediments. Recent seismicity also clearly delineates these faults and further attests to their activity. On land the San Miguel fault is active along its entire length and appears to be the seismically dominant zone in northern Baja California. It has produced several earthquakes of magnitude 6.0 to 6.8. No moderate or large-magnitude historical earthquake can be readily associated with the Agua Blanca fault, but earthquakes of magnitude 3 to 5 have

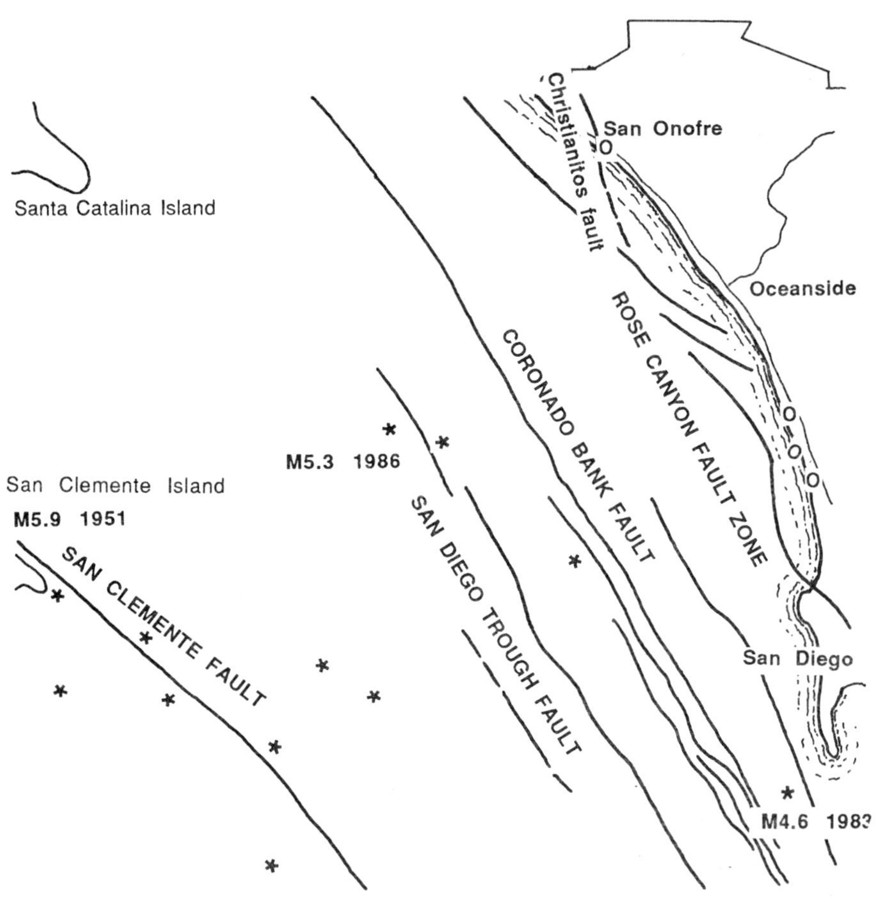

Map of San Diego County and offshore area showing active faults (labeled), earthquake epicenters (stars) from chapter VII, and fault localities described in chapter VIII (circles).

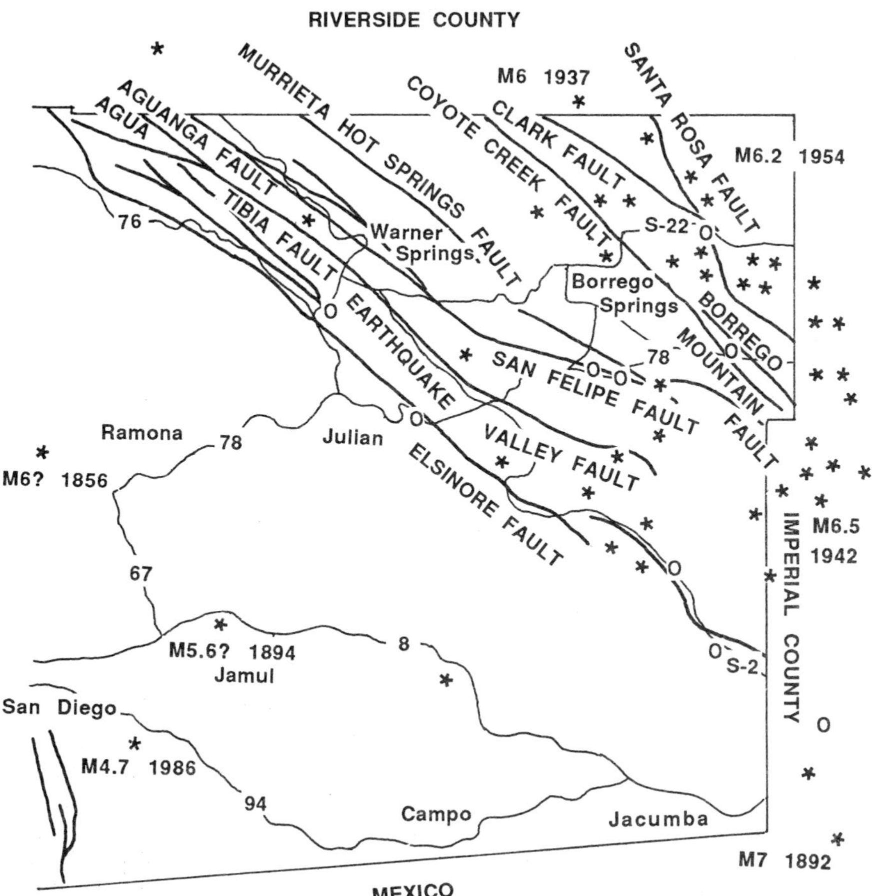
RIVERSIDE COUNTY
MURRIETA HOT SPRINGS FAULT
AGUANGA FAULT
AGUA
TIBIA FAULT
COYOTE CREEK FAULT
CLARK FAULT
SANTA ROSA FAULT
M6 1937
M6.2 1954
76
Warner Springs
S-22
Borrego Springs
BORREGO MOUNTAIN FAULT
78
EARTHQUAKE
SAN FELIPE FAULT
Ramona
78
Julian
VALLEY FAULT
M6? 1856
ELSINORE FAULT
67
IMPERIAL COUNTY
M6.5 1942
M5.6? 1894
8
O S-2
Jamul
San Diego
M4.7 1986
94
Campo
Jacumba
MEXICO
M7 1892

occurred along its offshore portion. The Coronado Bank fault zone also is characterized by a linear trend of magnitude 3 to 5 earthquakes offshore from San Diego. The July, 1986, quake of magnitude 5.3 was centered within the northern portion of the Coronado Bank fault zone west of Del Mar. The San Clemente fault zone also has high seismicity; the largest historic earthquake in that zone was magnitude 5.9 in 1951.

The only two earthquakes known to have caused significant damage in the coastal zone were those of 1800 and 1862. Intensity VII shaking associated with the 1800 quake is indicated by reports of cracked adobe walls both at San Diego and San Juan Capistrano. It is estimated that the quake must have had a magnitude of at least 6.5 to have caused such damage in both places. There are not enough records to locate an epicenter, but it presumably was in the coastal zone near or between the two missions. The epicenter of the 1862 quake must have been in or just west of San Diego, probably in either the Rose Canyon or Coronado Bank fault zones. Magnitude is estimated at about 6, and intensity of VII is indicated by damage reports at two locations in San Diego.

With these two exceptions earthquake activity has been low in the coastal zone during the past 200 years and especially during this century. From 1932 to 1982 the largest quake in the immediate vicinity of San Diego was one of magnitude 3.7 in 1964. Between 1983 and 1986, however, coastal San Diego County was struck by earthquakes of magnitude 4.6 (June 29, 1983), 4.3 (July 1, 1984), 4.0 (June 17, 1985), and 5.3 (July 13, 1986). Some 40 quakes of magnitude 3.5 or greater struck the broader San Diego area from 1984 through 1986, matching the number between 1944 and 1984. Though the significance of that flurry of quakes is not yet understood, similar suddenly increased activity has, in some cases, preceded a stronger quake.

Future earthquake estimates. On land the northwestern I2- to I8-mile segment of the Agua Blanca

fault is thought to be capable of an earthquake of M6.7 to
7.2 (Anderson, Agnew, and Rockwell, 1989). The average
recurrence interval for such quakes would be about 125 to
250 years.

Leighton and Associates (1983) estimated a maximum
probable earthquake on the Coronado Bank fault of
magnitude 5.8 to 6.2, with a recurrence interval of 100 to
200 years. Woodward-Clyde Consultants (1986)
suggested a magnitude range of 6.2 to 7.2, while Anderson,
Agnew, and Rockwell (1989) estimated a maximum
plausible magnitude range of 6.1 to 7.7 for earthquakes
rupturing multiple segments of this fault.

The maximum earthquake range for the San Diego Trough
fault has been estimated at magnitude 6.2 to 6.7
(Woodward-Clyde Consultants, 1986) and 6.1 to 7.7
(Anderson, Agnew, and Rockwell, 1989).

The poorly known San Clemente fault is thought to be
capable of earthquakes up to magnitude 7.7 (Woodward-
Clyde Consultants, 1986; Anderson, Agnew, and Rockwell,
1989).

ROSE CANYON FAULT ZONE

Fault descriptions -- Rose Canyon fault. Dominant
among the County's onshore coastal faults is the one in Rose
Canyon. This fault crosses San Diego from south-southeast
to north-northwest, passing from San Diego Bay along the
east shore of Mission Bay, through Rose Canyon, and out to
sea at La Jolla Shores. It extends beneath the sea at least
25 miles northward to the sea floor west of Oceanside, and
it may continue on to connect with the Newport-Inglewood
fault -- site of the 1933 Long Beach earthquake. It is
parallel to the San Andreas and other major faults of the
southern California region, and rocks west of the Rose
Canyon fault are similarly plowing northward past rocks
on its eastern side. Sandstones of the San Diego Formation,
for example, are found high on the south flank of Mount

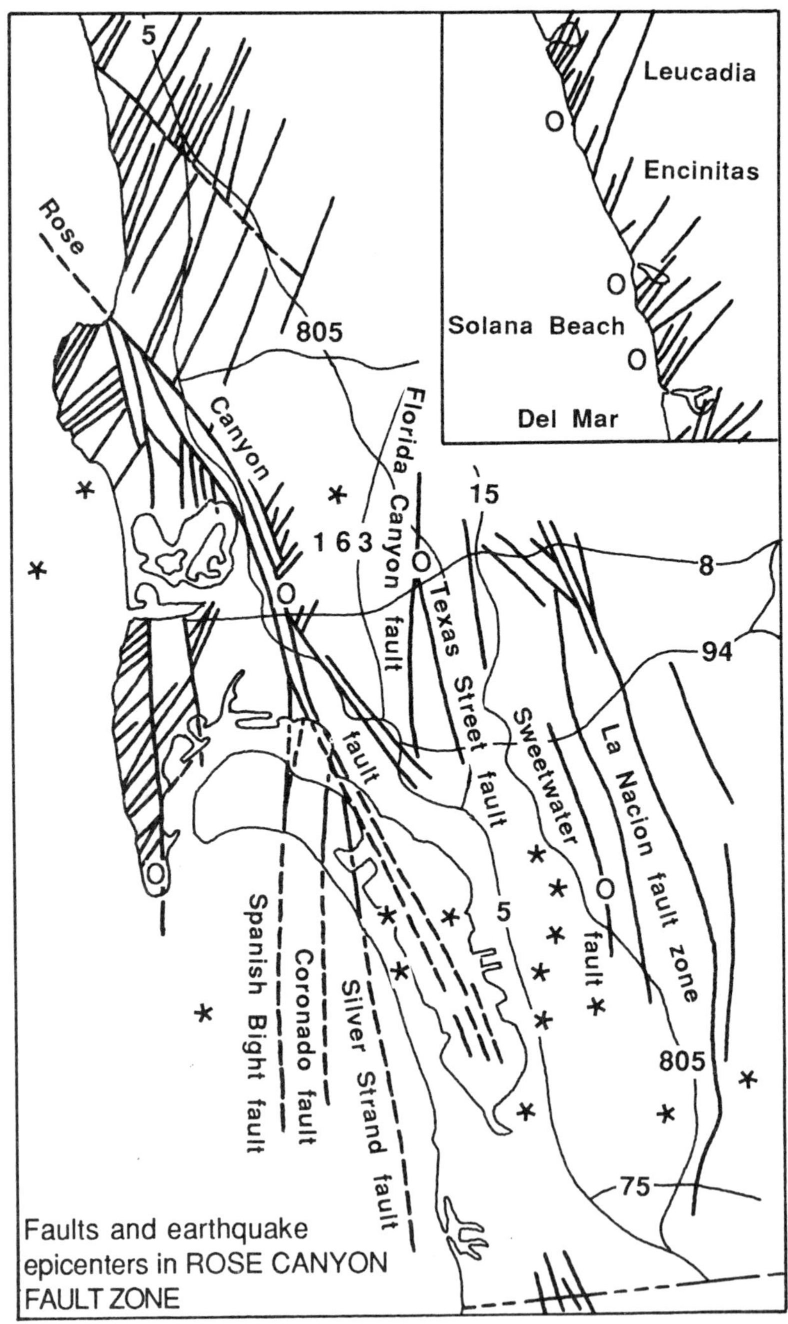

5
Rose
805
Canyon
Florida Canyon fault
Leucadia
Encinitas
Solana Beach
Del Mar
15
8
94
163
Texas Street fault
Sweetwater fault
La Nacion fault zone
fault
5
Spanish Bight fault
Coronado fault
Silver Strand fault
805
75
Faults and earthquake epicenters in ROSE CANYON FAULT ZONE

Soledad west of the fault but are restricted on its east side
to the south wall of Mission Valley. Though it apparently is
much less active and powerful than its notorious relatives,
these features reveal the Rose Canyon fault as a member of
that zone of fractures that mark the boundary between the
Pacific and American plates.

Mount Soledad and south La Jolla thus are slipping seaward
past La Jolla Shores, while Point Loma follows them past
downtown San Diego. The illustration shows where the
bend in the fault at Ardath Road has further caused the
western block to ride upward there. Mount Soledad has
thus been thrust above the mesa, and Rose Canyon has been
deflected southward and become deeply incised behind the
intruding mountain. The main fault strand crosses the
steep slopes south of Ardath Road, where the uplifted block
towers over its neighbor to the east and north. Farther
south, on the other hand, Mission Bay and San Diego Bay
have subsided below the level of the mesa east of that
stretch of the fault, which traverses the steep slope above
those shores. These relationships can be seen on a clear
day from the Easter Cross on Mount Soledad. The fault's
continued southward course is uncertain where covered by

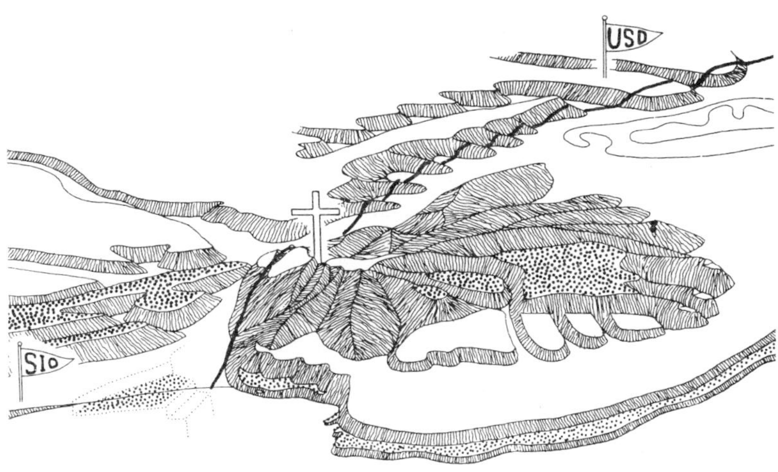

Mount Soledad

the city and the water, and the zone appears to die out under San Diego Bay, with its motions passing on to the offshore faults to the west.

Rocks and faults are poorly exposed in San Diego because they are covered by streets, parking lots, buildings, lawns, and shrubbery. In spite of its prominence, therefore, the Rose Canyon fault is not easily seen. One of its several strands, however, is exposed behind the left-center-field fence of the Pony League baseball diamond in Tecolote Park's south corner. Half-million-year-old (Pleistocene) conglomerate is on the right side of this fault, with light-colored 50 million-year-old (Eocene) sandstone of the Scripps Formation to the left.

Incidentally, this exposure is being made more visible and will be part of an earthquake hazard exhibit in the visitor center that is planned for construction in 1990 in Tecolote Canyon Natural Reserve. This exhibit will include the Rose Canyon fault model described in chapter IX.

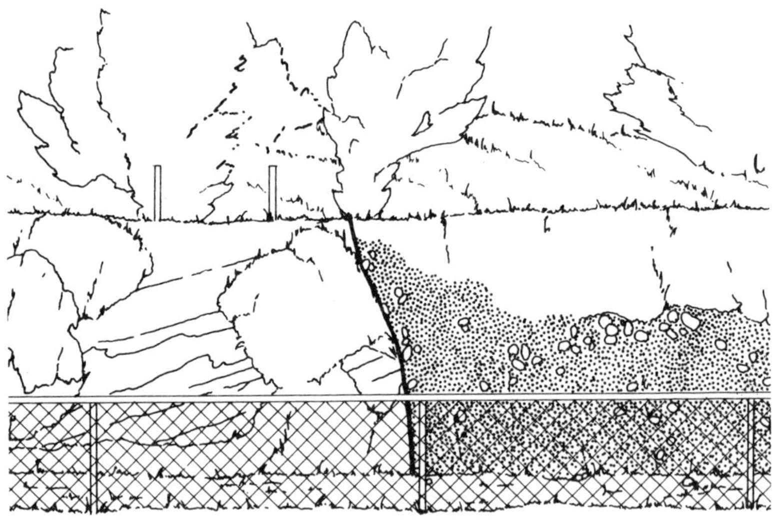

ROSE CANYON FAULT in Tecolote Park

The youngest features that are known to have been dismembered by the Rose Canyon fault, and the ones incidentally that most clearly reveal its nature, are a series of Pleistocene marine terraces at La Jolla Bay. These are flat benches, like the one exposed by low tides at Bird Rock or the tip of Point Loma. They are carved by the sea and may later be lifted above the shore where the crust is buckled upward by folds or faults. For example, the grassy surface of the park at La Jolla Cove (fine stipple in the illustration) and the higher surface under downtown La Jolla have been pushed respectively 80 feet and 200 feet upward above their subterranean levels beneath La Jolla Shores on the other side of the Rose Canyon fault. Those terraces are only about 105,000 and 120,000 years old, so the fault has been busy here in that short span of time. Another terrace that is more than half a million years old has become even more disjointed because of the fault. It is conspicuous in the Mount Soledad illustration (coarser stipple) at elevations of 340 feet north and east of Ardath Road and 625 feet across the Rose Canyon fault on Mount Soledad.

The San Diego fault map shows that nearly all other large-scale fractures in the city diverge from the Rose Canyon fault like the branches of a tree, each extending only a short distance before dying out. Most of these branches, however, are oriented at clockwise angles of approximately 30 to the trunk, so those east of the main fault extend toward the north while those on its western side reach to the south. This distinctive pattern is characteristic of fault zones, like the San Andreas and its partners in the southern California region, along which rocks viewed across the trunk fault have moved to the right. The specific term "right-slip fault" is used by geologists to distinguish this category of faults.

None of the many faults in this extensive branching system are active. This is clear from the fact that none of them cut the 105,000- and 120,000-year-old terraces. It is also characteristic of right-slip (and left-slip) fault zones that they originate as extensive sets of faults crossing the zone and that those faults become inactive when the main fault itself later develops. Many of the branch faults are normal faults, in which the fracture plane is sloping and the overhanging block has slipped down relative to the underlying one. This relationship can be seen in the illustration of the Florida Canyon fault, as well as in several subsequent illustrations.

Following are brief accounts of several of the principal faults that branch from the Rose Canyon.

Florida Canyon fault. In spite of generally poor exposure of rocks in San Diego, many of the principal secondary faults, or at least their topographic effects, can be seen. A good example is the Florida Canyon fault, which achieved some notoriety because of its proximity to the Naval Hospital in Balboa Park. Adams Avenue, El Cajon Boulevard, University Avenue, Morley Field Drive, and Zoo Place all plunge steeply just east of Park Boulevard. That abrupt descent marks the fault scarp -- the eroded face of the uplifted block -- where the mesa surface east of the fault has subsided as much as 65 feet below its level to the west. The illustration shows this topographic relationship, as well as the corresponding displacement of red terrace sandstones that blanket the mesa. Florida Canyon has been carved into the gently sloping mesa by the eroding action of surface water that begins flowing seaward but is trapped behind the uplifted block and diverted toward the south. Even this terrace is less than one million years old, showing that the Florida Canyon fault has been active within that interval.

The best exposure of the Florida Canyon fault is in the cut bank along Friars Road just east of the Conrock quarry and Stadium Way. Partially obscured by brush at the eastern

FLORIDA CANYON (left) and TEXAS STREET FAULTS

FLORIDA CANYON FAULT at Friars Road

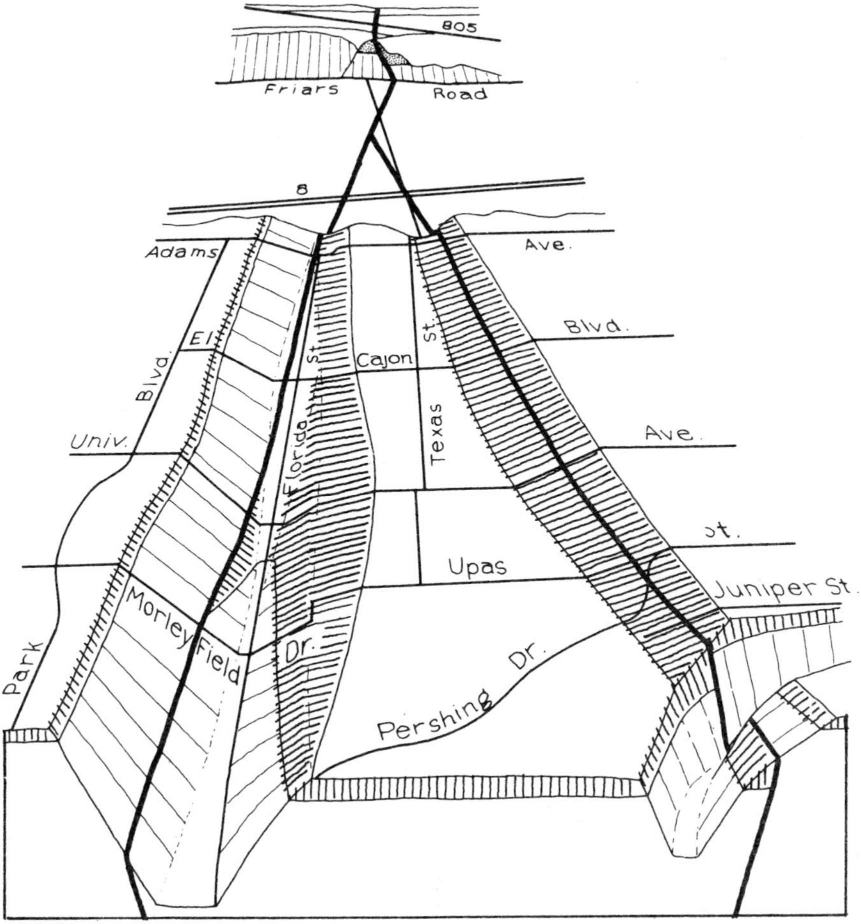

805
Friars Road
8
Adams Ave.
El Blvd.
Univ. Blvd.
Cajon St.
Florida St.
Texas St.
Ave.
St.
Park
Morley Field
Dr.
Upas
Juniper St.
Pershing Dr.

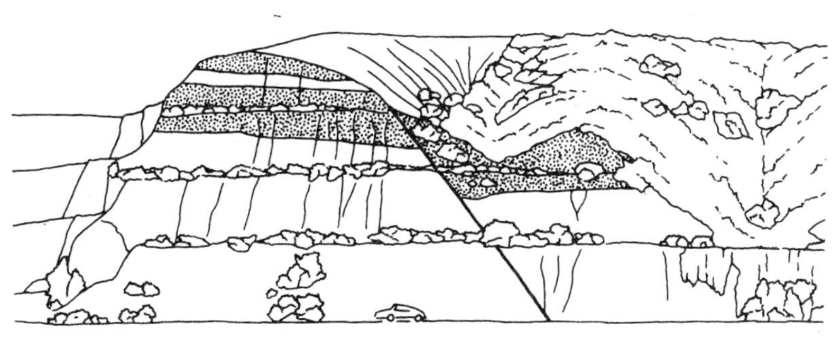

end of the high road cut, the fault plane slopes down toward the right. A dark brown zone (stippled in the illustration) within the Stadium Conglomerate can be seen to be faulted down to a lower level in the overhanging eastern block. Parking is not possible on Friars Road, so this exposure is best seen from Stadium Way and also from the Adams Avenue bridge over Texas Street. The Florida Canyon fault also can be seen in a low road cut on the north side of Morley Field Drive just west of Florida Drive.

Texas Street fault. The terrace that is cut by the Florida Canyon fault is broken by three other faults on San Diego Mesa. Though the fault plane itself is nowhere clearly visible, the Texas Street fault is conspicuous by its topographic expression. As shown in the drawing one can climb its scarp by traveling eastward from Texas Street on Adams Avenue, El Cajon Boulevard, or University Avenue. On Upas Street the scarp is at Pershing Avenue, and it forms the steep slope along Pershing Drive at the northeast corner of Balboa Park and on Juniper and Grape streets east of the Park.

Fortieth Street fault. The Fortieth Street fault is like the one in Florida Canyon in sloping eastward and having the west side higher. Though its surface expression is not so evident to the traveler of city streets, westward-flowing surface water has similarly been diverted to north and south, carving small canyons that descend in those directions, as along Ward Road.

La Nacion fault. The several strands of the La Nacion fault system mimic the geometry of the Texas Street fault. They slope in most places toward the west, and the mesa surface is lower in that direction. In this case, in fact, it drops down in several steps which one ascends in traveling eastward along El Cajon Boulevard from Euclid Avenue past 54th Street. As at Florida Canyon and Texas Street these faults cut the veneer of red terrace sandstone that covers the mesa.

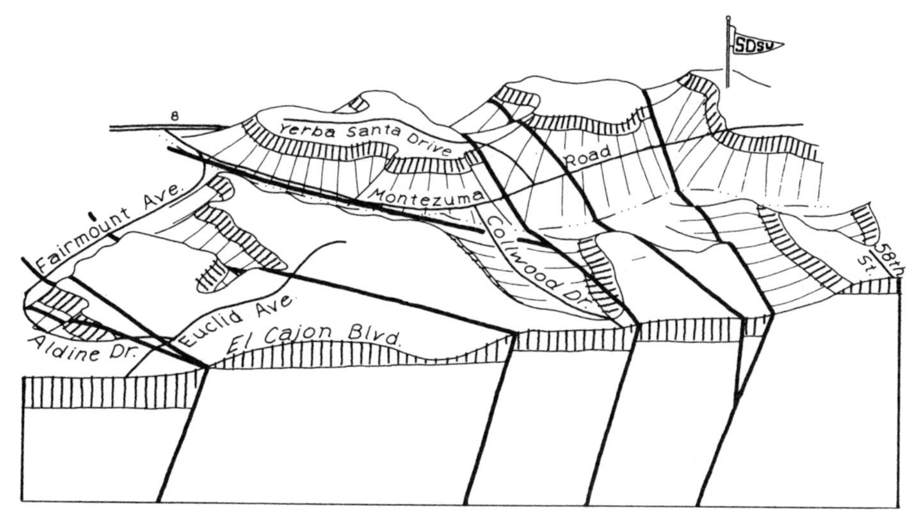

LA NACION FAULT ZONE

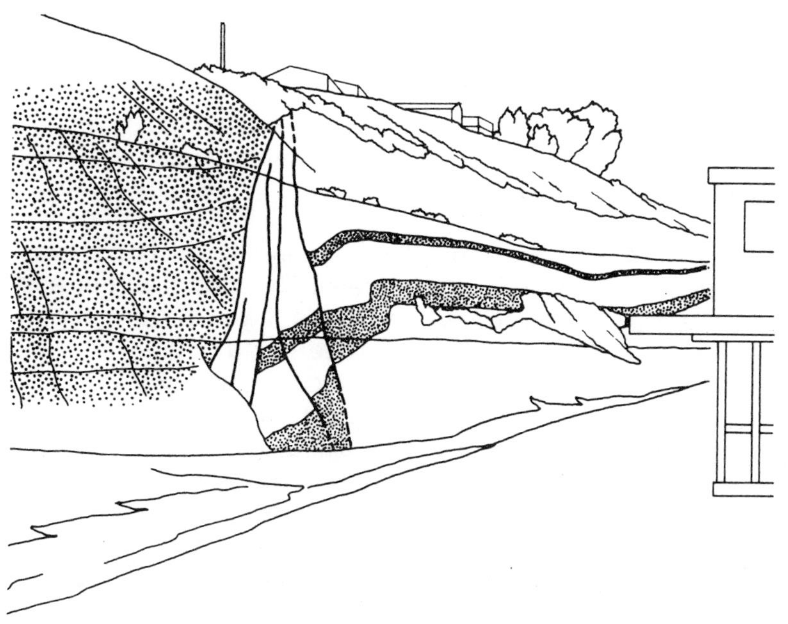

SWEETWATER FAULT at Twin Hills Youth Park

The La Nacion fault zone extends in a broad swath from this area southward through National City and Chula Vista to the Mexican border, and it is exposed in many places. National City has one outstanding exposure in the northern bluff overlooking Twin Hills Youth Park on Calle Abajo near Valley Road and Sweetwater Road. The illustration shows how red-brown Pleistocene sandstone and conglomerate in the overhanging block have dropped down at least 50 feet on the left side of the Sweetwater fault to abut against lighter-colored sandstone of the several-million-year-old (Pliocene) San Diego Formation.

Faults on Point Loma. Another series of prominent secondary faults branches from the Rose Canyon zone toward the south and southwest. Those under San Diego Bay and the ocean west of Silver Strand were detected from the echoes that offset beds reflect back to oceanographic research ships from sound waves they have transmitted. Associated faults on land are not clearly exposed, but several of them are topographically conspicuous. The valley occupied by Nimitz Boulevard, for example, has been eroded along a fault that pushed up the high area on its eastern side. That small plateau also is bounded on its northwestern edge by another fault that produced the steep bluffs along Worden Street.

The Point Loma peninsula itself apparently exists because of uplift north of faults that parallel Rosecrans Street and west of others that follow the Point's lower eastern shore. A prominent fault in the latter zone cuts through approximately 70 million-year-old (Cretaceous) sandstones of the Cabrillo Formation. It can be seen near the eastern end of the high sea cliff at the extreme southern end of the Point, but a more accessible if slightly less clear exposure is in the north wall of the first canyon entered by the Bayside Trail in Cabrillo National Monument. At the latter site a shaly bed is conspicuously lower on the bay side of the fault.

ROSE CANYON FAULT ZONE. Fault on Bayside Trail,
Cabrillo National Monument, San Diego

Faults near Mission Bay. Several less-prominent,
unnamed faults also can be seen clearly in readily
accessible exposures. One of these is in the high cut on the
east side of Soledad Mountain Road just north of the traffic
light at the end of Beryl Street. (This is a busy street;
watch for traffic here.) Red and gray sandstones of a
Pleistocene terrace deposit (stippled in the illustration)
in the overhanging block on the left are bent slightly where
faulted down against tan and gray Pliocene sandstone of the
San Diego Formation. Above the Pliocene beds the
Pleistocene rocks are hidden beneath the houses at the top
of the hill, where they have been heaved 45 feet above
their level on the other side of the fault.

ROSE CANYON FAULT ZONE. Fault on Soledad Mountain Road, San Diego

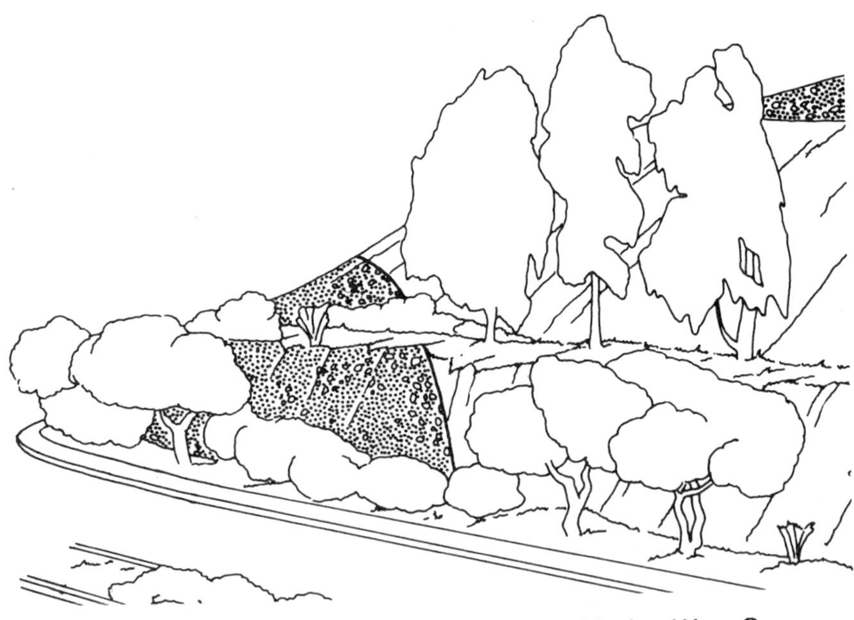

ROSE CANYON FAULT ZONE. Fault on Marion Way, San Diego

Another, lesser branch fault is exposed in the bank along
the western entrance road (Marion Way) to the University
of San Diego, near the upper end of the bend halfway up the
hill. Here brown conglomerate of a Pleistocene terrace
deposit (stippled in the illustration) in the overhanging
block on the left has been faulted down against tan Eocene
sandstone of the Scripps Formation. Above the Eocene
rocks the same Pleistocene beds lie 50 feet higher at the
top of the hill. The terraces offset by both of the preceding
faults are intermediate in age, perhaps as young as one-
half million years, so the age of faulting here must be at
least that young.

Faults from Del Mar to Oceanside. Many of the
presumably inactive northeast-trending faults of the Rose
Canyon zone can be seen along the coast, especially in sea
cliff exposures, from Del Mar to beyond Oceanside. There
are fine fault outcrops in the bluffs south of Flat Rock in
Torrey Pines Park, where the Rose canyon fault lies just
offshore. Exposures are poor from Torrey Pines through
Del Mar, but the coastal bluffs from south Solana Beach all
the way to San Elijo Lagoon are sliced by faults in many
places. There are especially good exposures on the beach
just south of Via de la Valle and south of the parking lot at
Cardiff State Beach at the north end of Solana Beach. In
both these places many faults can be clearly seen to cut

ROSE CANYON FAULT ZONE. Faults in sea cliffs south of
Via de la Valle, Solana Beach

through the bedrock sandstones but not through the overlying soft terrace sands, showing that these faults have not been active during the past 120,000 years.

ROSE CANYON FAULT ZONE. Faults in sea cliffs at Cardiff State Beach, Solana Beach

ROSE CANYON FAULT ZONE. Faults in sea cliffs at Leucadia State Beach, Leucadia

For several miles north of San Elijo Lagoon exposures also are poor, but faults can be seen at several places in Leucadia. They are sparse at Moonlight State Beach, but there is a small cluster just beyond the stairs 200 yards south of the parking lot. A better place is below the parking lot at Leucadia State Beach at the foot of Leucadia Boulevard, where faults cut the sandstones and shales both north and south of the base of the beach path.

North of Leucadia the Rose Canyon fault is farther offshore, and few of the northeast-trending faults reach the coast. A prominent exception is the Christianitos fault, which can be seen in the sea cliffs at San Onofre State Beach, one mile south of the San Onofre Nuclear Generating Station. This fault, which was of great concern in the siting of the power plant there, appears to be another of the northeast-trending faults of the Rose Canyon zone. It diverges in a clockwise direction from the offshore fault zones, and its failure to cut the young terraces shows it to have been inactive for the past 125,000 years.

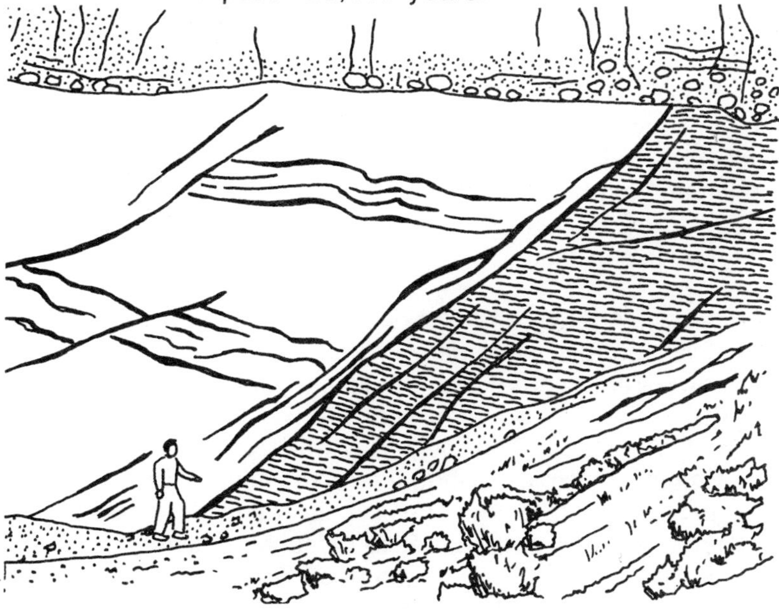

CHRISTIANITOS FAULT, San Onofre State Beach

Earthquake history. The historic earthquake record of the Rose Canyon fault zone is ambiguous because of the uncertainty of epicentral locations of the 1800 and 1862 quakes. Both produced intensities of VI to VII in San Diego and probably were located either on the Rose Canyon fault (with magnitudes of approximately 6.5 and 6.0) or on the Coronado Bank fault. The left bend in the Rose Canyon fault north of Mount Soledad may be locking the fault, perhaps accounting for its low seismic profile during the past century or more. The general lack of earthquake activity in the coastal zone during the past two centuries was broken onshore as well as offshore in the 1980's, as the June 17, 1985, (3.9, 4.0, and 3.9) and October 28, 1986, (4.7) quakes were centered under the city.

Future earthquake estimates. As information has accumulated during the 1980's, earthquake risk associated with the Rose Canyon zone has been receiving increasing attention. For example, an evaluation of San Diego's seismic sources, which was done as part of a liquefaction study by Woodward-Clyde Consultants (1986), suggested maximum magnitude ranges of 6.2 to 7.2 for the Rose Canyon fault and 6.2 for the La Nacion fault. Other faults in the zone from the La Nacion to east Point Loma can be expected to be at least comparable. Wesnousky (1986) estimated expected earthquakes of magnitude 7.1 on the Rose Canyon fault (30-mile segment length) and 6.6 on the La Nacion fault.

Anderson, Agnew, and Rockwell (1989) have evaluated seismic potential according to probable rupture length of eight different segments of the Rose Canyon fault. Rupture of the entire well-documented 35-mile length of the fault from Oceanside to San Diego Bay would produce a maximum plausible earthquake of magnitude 6.9. Magnitude 6.4 to 6.6 quakes would result from partial ruptures from La Jolla either to Oceanside or through San Diego Bay. The 40-mile segment off Camp Pendleton could be capable of a quake of magnitude 7.0. In addition, the 12-mile-long La Nacion fault is suggested to be capable of a magnitude 6.5 earthquake, and the same presumably is true of any of the other faults across that zone from there to Point Loma.

New evidence has recently been brought to light by Tom Rockwell, who found that 1929 aerial photos show 100 feet of offset of the modern creek bed in Rose Canyon along the Rose Canyon fault. Most of that slip must have occurred during the past few thousand years, showing clearly that the Rose Canyon is an active fault.

According to California Division of Mines and Geology seismologist Michael Reichle, a magnitude 6.5 earthquake on the Rose Canyon fault could produce shaking intensities as high as MM VIII or IX in San Diego valleys and along bays. Earthquakes in the ranges given above for any of the onshore or nearshore San Diego faults could subject the coastal zone to shaking intensities of VII (at epicenter of magnitude 6 earthquakes) or even as high as IX to X (at epicenter of magnitude 7 quakes).

ELSINORE FAULT ZONE

Fault descriptions. The Elsinore fault zone crosses eastern San Diego County on its 120-mile path from Los Angeles County to the Mexican border. From Lake Elsinore it passes near Temecula and up Wolf Valley to enter San Diego County where S-16 crosses the crest north of Pala. The base of the steep southwestern flank of the Agua Tibia-Palomar Mountain ridge marks the passage of the fault toward Lake Henshaw. The alignment of the lake's straight southwestern shoreline with the narrow, steep-walled valleys occupied by route 76 northward from the lake and route 79 to the south reveal the fault's path here. Two separate parallel strands of the fault cross the lake itself, and one of them can be seen in a road-cut exposure on route 79 a half mile north of the junction with 76 at the south end of the lake. (I am reluctant to recommend this exposure.   The cut is narrow and traffic is fast here; park well off the road west of the cut and keep out of the way of traffic.) Recent activity on this fault is shown by its

ELSINORE FAULT in road cut south of Lake Henshaw

cutting of these Pleistocene beds, at most a few million years old.

Three to four miles down route 79 toward Santa Ysabel the road bends abruptly to the right. Here the fault diverges left from the road and continues straight on up the linear valley that can be seen to the southeast beyond the bend. It then goes over a pass above the apple orchards on Farmers Road north of Julian and continues on down Banner Grade. On a road map the fault's path can be seen in the alignment of Banner Grade, routes 76 and 79 near Lake Henshaw, and the northern few miles of S-16.

There are excellent views of the Elsinore fault from any of several turnouts on Banner Grade between about 1.5 and 3 miles above the Banner Store. (Park well off the road on the canyon side and watch for traffic.) From these turnouts one can look across the canyon to where the fault line is marked by several distinct benches, by lines of taller vegetation where water comes up along the fault, and

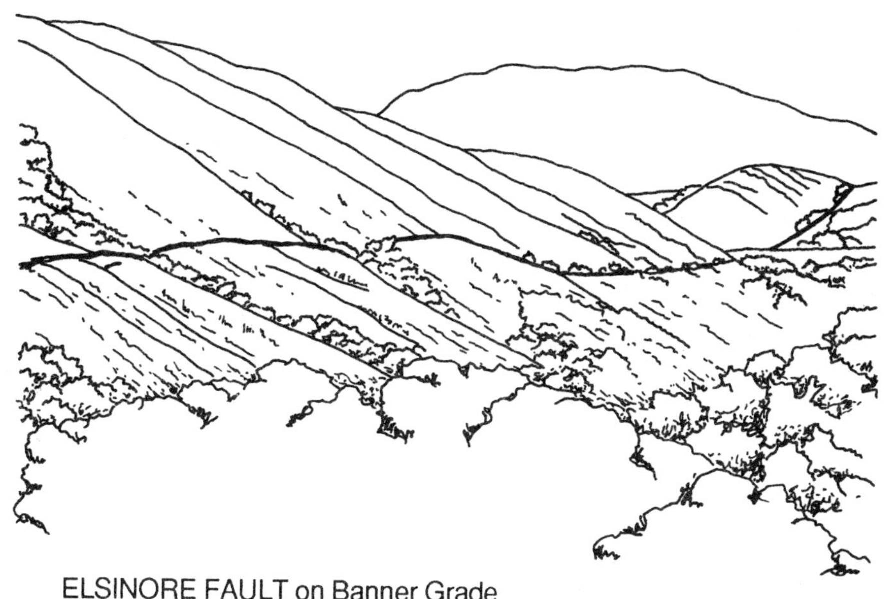

ELSINORE FAULT on Banner Grade

by steeper slopes (fault scarps) in the lower end of the canyon. From directly across the canyon the two most prominent benches can be seen to be the displaced lower ends of ridges, which have moved to the left (northward) on the near side of the fault. The gullies between these ridges also can be seen to have been displaced to the left by movement on the fault.

From Banner the fault continues southeastward up the straight canyon and on over the saddle west of Granite Mountain. From there it descends into the next valley and crosses County Road S-2 in the bend at the base of the grade. The patchwork of dark and light rocks on the slopes north of the Butterfield Ranch is the result of faulting there. The fault zone crosses the road again into the valley south of the Vallecito Stage Station, then bends left over the ridge to Agua Caliente Springs. This bend marks the boundary between two of the major segments of the fault zone.

The hot springs at both Vallecito and Agua Caliente result from hot water moving upward through the fractured rocks in the fault zone. At Agua Caliente there are wide zones of crushed rock produced by grinding along the fault. Much of the crushed zone is filled with clay produced by rapid decomposition of feldspars in the granitic rocks partly because of the presence of the hot water. The zone of faulting and crushing is some hundred yards wide at Agua Caliente. Several principal strands are shown in the illustration, and at the base of the slope in the foreground are pinnacles of crushed and clay-seamed rock.

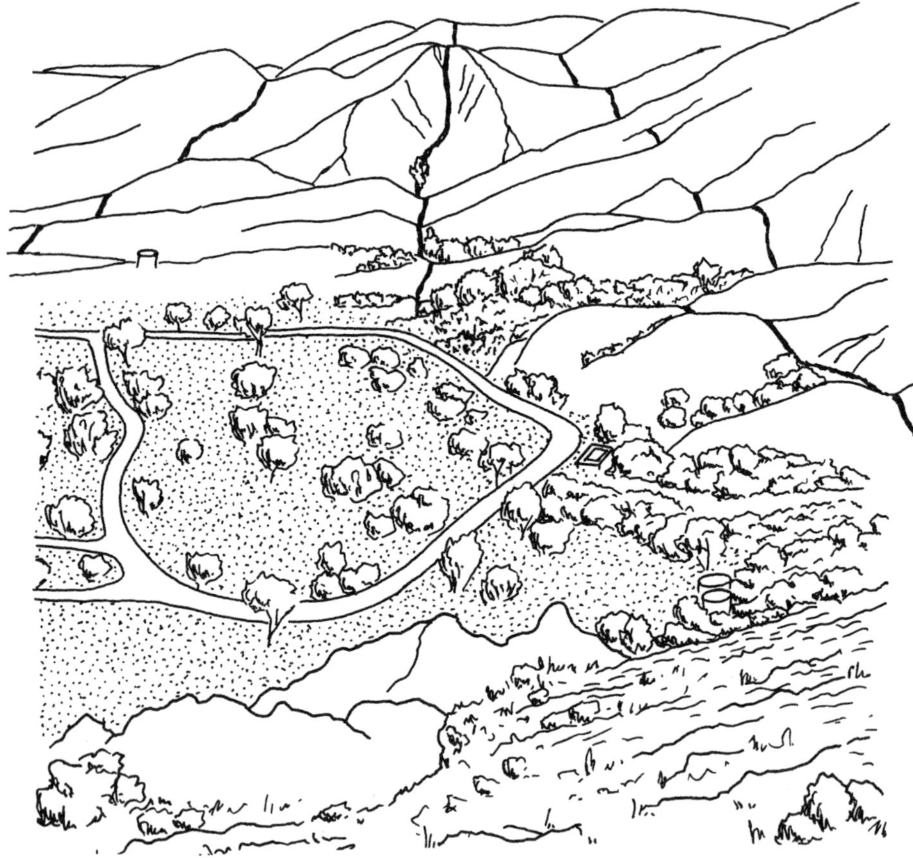

ELSINORE FAULT seen from Desert View Overlook at Agua Caliente Springs

ELSINORE FAULT from overlook at Canyon Sin Nombre, north of Ocotillo. Elsinore fault crosses view at base of slope

Southeastward from Agua Caliente the fault follows the base of the Sierra Blanca Mountains west of the road, and water in the Mountain Palm Springs area flows up along the fault plane. The fault zone crosses the road again in Carrizo Wash and continues on toward Imperial County along the base of the southwestern flank of the Coyote Mountains. Deformation caused by the Elsinore and some northeast-trending branch faults is clearly visible at the head of Canyon Sin Nombre, and it can be seen even from the overlook road just off S-2 southeast of Sweeney Pass and about four miles from the County line.

A mosaic of colored rocks along the whole length of the Coyote Mountains reveals the intense faulting that has occurred there, and the Elsinore fault follows the base of the slope all the way to Ocotillo, where it passes on under the sandy desert floor. There are excellent views of the fault at the mouth of Fossil Canyon, which can be reached by a short gravel road directly north of Ocotillo.

ELSINORE FAULT ZONE at mouth of Fossil Canyon

Earthquake history and future earthquake estimates. The maximum probable earthquake for the Elsinore fault zone as a whole has been estimated at magnitude 6.9 to 7.3, with a recurrence interval of 60 years (McEuen and Pinckney, 1972; Woodward-Clyde Consultants, 1986). Recent and on-going studies also have suggested characteristic earthquakes for individual fault segments in the range of magnitude 6 to 7, as described below.

The northwesternmost of four segments (segment A) extends approximately 45 miles from Whittier, in Los Angeles County, to the prominent right step at Lake Elsinore, in Riverside County. The 1910 magnitude 6 earthquake in Temescal Valley occurred on this segment and caused intensity V to VI shaking in northern San Diego County. Wesnousky (1986) suggested an expected earthquake magnitude here of 7.3. Trench studies on this segment have revealed the occurrence of five earthquakes during the past few thousand years. The maximum groundbreaking earthquake recurrence interval of 250 years, combined with a moderate estimated slip rate of 4 to 7 millimeters per year, suggests that the characteristic quake on this segment is magnitude 6 to 7.

The next segment to the southwest (segment B) extends some 30 miles from Elsinore to the sharp bend at Pauma Valley in northern San Diego County. Lack of major historical earthquakes on this segment, combined with a significant slip rate, suggests that magnitude 7 earthquakes may be characteristic here. Anderson, Agnew, and Rockwell (1989) suggested that the maximum plausible earthquake on this segment is magnitude 6.8, and Wesnousky (1986) gave magnitude 7.1 as the maximum expected quake.

The third segment toward the southeast (segment C) extends some 40 miles from Pauma Valley to the prominent bend at Vallecito. Though no data on earthquake history have yet been obtained in this segment, Anderson, Agnew, and Rockwell (1989) used segment length to estimate a slip rate of 5 to 6 millimeters per year and a maximum expected earthquake of magnitude 7.1. Wesnousky (1986) suggested an expected earthquake magnitude of 7.2.

The southeasternmost segment (segment D) of the Elsinore fault itself extends the last 40 miles to the Mexican border south of Plaster City. Measurements of offset stream channels reveal the occurrence of six prehistoric earthquakes. The amount of horizontal slip per quake ranges from 3 to 6 feet, suggesting magnitude 6.5 to 7 quakes. A recurrence interval of about 350 years has been estimated for these displacements, with the last one being prehistoric. Anderson, Agnew, and Rockwell (1989) estimated a moderate slip rate of 3.5 to 5.5 millimeters per year and an expected quake of magnitude 7, and Wesnousky (1986) suggested a maximum magnitude of 7.2.

The final segment in this fault zone is the Laguna Salada fault, which extends 35 miles beyond the Mexican border. The last earthquake, probably the historic magnitude 6.7 to 7.3 quake in 1892, produced vertical scarps up to 12 feet high (see chapter I) and ruptured for a length of at least 12 miles. Intensity VII to VIII shaking occurred in the southeastern corner of the County, and the San Diego coastal zone had intensities of VI to VII.

FAULTS NORTHEAST OF THE ELSINORE FAULT

Fault descriptions. A 12-mile-wide zone on the northeast flank of the Elsinore fault is occupied by three major active fault zones. From southwest to northeast these are the Agua Tibia-Earthquake Valley zone, the Aguanga-San Felipe zone, and the Murrieta Hot Springs fault.

The Agua Tibia fault crosses Palomar Mountain just west of High Point and along the east flank of Barker Valley to cross the middle of the Lake Henshaw valley. From there it follows S-2 past San Felipe, where it passes into the Earthquake Valley fault. Movement on this zone has put Earthquake and San Felipe valleys under the range of hills on their east flank, along the base of which the faults lie. They cross state route 78 without good exposure just east of Scissors Crossing, and the zone dies out in the Pinyon Mountains 20 miles to the southeast.

The Aguanga fault lies along the northeast flank of the Agua Tibia-Palomar Mountains and is responsible for the steep slopes above Aguanga and Oak Grove. It also passes through Warner Springs, where hot water rises to the surface through fractured rocks along several separate strands of this fault zone. Several miles of S-22 west of Ranchita are lined with road cuts in which the rocks are veined and fractured in the zone of this fault, which crosses the road but is not exposed near the first bend west of town.

SAN FELIPE FAULT viewed from turnout on highway 78 a mile and a quarter east of Tamarisk Grove junction

Beyond a bend in the Grapevine Hills the Aguanga fault connects with the San Felipe fault. From a large highway 78 turnout on a low crest a mile and a quarter east of the Tamarisk Grove Campground junction (watch for traffic, which travels fast on this road) the crushed and bleached rocks of the San Felipe fault zone can be seen crossing S-3 above the campground. (S-3 is the heavy diagonal line in the illustration.) Five miles east of the junction there is another turnout on route 78 at the San Felipe Creek narrows, where you may find printed guides to the Narrows Earth Trail. This trail is entirely within the fault zone, and there are fine exposures of faults at stops 4 and 6 and of crushed and fractured rock throughout the zone. At stops 6 and 7 the main fault strand puts light-colored granitic rock (to the south) against darker, banded metamorphic rock. Beyond here this zone breaks up into scattered faults, including some of those seen in Split Mountain Gorge.

The Murrieta Hot Springs fault is in the remote mountains northeast of route 79 in the Aguanga-Warm Springs area. It becomes discontinuous southeastward, where it merges with the San Felipe fault zone near the San Felipe Creek narrows.

Future earthquake estimates. There are no published earthquake risk studies of any of the faults in this zone, but all are thought to be probably active with low slip rates of no more than a few millimeters a year. As individual segment lengths are comparable to those of the Elsinore fault, the faults in this zone presumably are capable of earthquakes in the same magnitude range but with long recurrence intervals.

SAN JACINTO FAULT ZONE

Fault descriptions. The San Jacinto zone is a complex fault system that is some 6 miles wide and 150 miles long, extending from its junction with the San Andreas zone near

SAN FELIPE FAULT at stops 6 and 7 on the Narrows Earth
Trail, at turnout on highway 78 five miles east of
Tamarisk Grove junction

Wrightwood to the northern edge of the Gulf of California. The zone passes diagonally across the northeastern corner of San Diego County, where it includes the Coyote Creek, Clark, San Felipe Hills, Borrego Mountain, and many smaller, unnamed faults. Among nine segments that have been distinguished along the entire zone, the 20- to 25-mile-long Coyote Creek segment, including the Clark fault, occupies the northern half of the San Diego County part. The Anza gap segment, 20 miles long, adjoins it to the northwest, extending from the bend north of Borrego Springs into Riverside County. Southwest of the Coyote Creek segment the Borrego Mountain segment occupies the next 18 miles, as far as the Imperial County line. The next segment beyond that is occupied by the parallel strands of the 15-mile-long Superstition Mountain fault and the 12-mile-long Superstition Hills fault.

The Coyote Creek fault lies under desert sand coming out of the mouth of Coyote Canyon north of Borrego Springs, but it is exposed in the rocks of the northeast wall in the less accessible upper canyon. To the south the Coyote Creek fault is aligned with the Borrego Mountain fault, which passes through Borrego Mountain and crosses highway 78 at Ocotillo Wells. The epicenter of the 1968 earthquake (magnitude 6.5) was just north of Ocotillo Wells; the fault rupture extended across the highway just east of the road junction, and maximum slip near there was 15 inches.

Among the many branch faults that cross Borrego Mountain, probably the most accessible one is in Hawk Canyon. This is two miles up the Buttes Pass Road, which goes north from highway 78 a mile and a quarter east of the Borrego Springs Road. There are good views of the fault both at the head of the canyon and looking into its mouth from the Buttes Canyon Road.

East of the Coyote Creek fault the Clark fault passes southeastward out of Riverside County. It is buried under the sand across the floor of Clark Valley, but where it is joined by the Santa Rosa fault along S-22 north of the

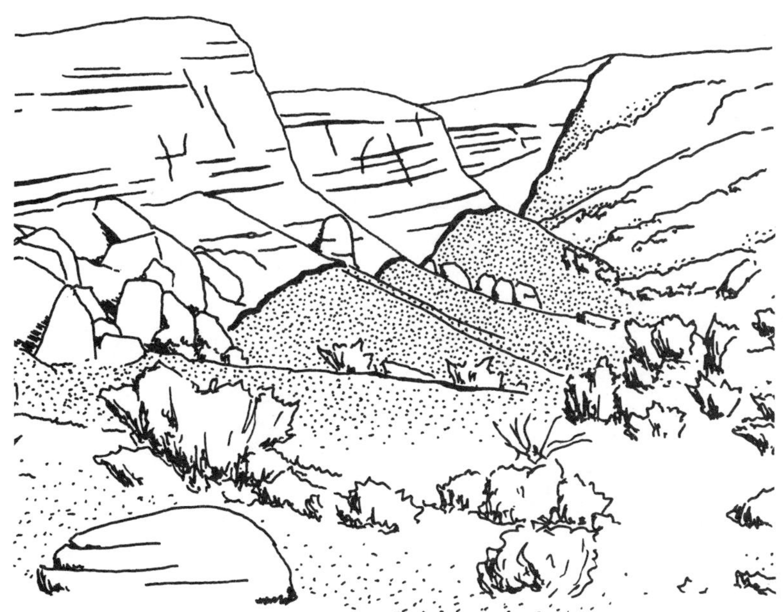

BORREGO MOUNTAIN FAULT ZONE. Fault at head of Hawk
Canyon

Borrego Badlands there are dramatic views of the effects of
recent fault activity. Several prominent cliffs (fault
scarps) in the alluvial fans mark places where recent
earthquakes have displaced these young deposits. Though it
is not visible from the road, the northeast slope of the
raised fault block just a quarter mile from the road at the
Truckhaven Trail turnoff near milepost 32 is a nearly
vertical straight scarp 150 feet high and a mile and a half
long.

Two miles to the west at milepost 30 there is a geologic
exhibit at a turnout on the south side of the road. The view
from here up Clark Valley includes two low sandy hills --
one near and one far -- that were formed by a buckling up
of the desert floor by movements along the Clark fault.
From this junction of the Clark and Santa Rosa faults the
San Felipe Hills fault passes out of the badlands south of the
road and on into Imperial County.

Junction of CLARK and SANTA ROSA FAULTS viewed from top of bluff on south side of S-22 at turnout west of milepost 34

Earthquake history. The San Jacinto fault is the most active of the southern California plate boundary zones. It has had 13 earthquakes of magnitude near or greater than 6.0 between 1899 and 1987. During that time it has produced more earthquakes in that range than any fault of comparable length in California.

From this short historical record it appears that earthquakes of magnitude 6 to 7 are characteristic of the entire San Jacinto zone and that these earthquakes occur at an average interval throughout the zone of 8 to 12 years. The Anza gap segment, just north of San Diego County, has had no earthquakes since the 1890's at least, so it presumably poses a more imminent seismic threat than do the others.

MAJOR SAN JACINTO FAULT ZONE EARTHQUAKES

DATE	EPICENTER	MAGNITUDE	INTENSITY
1899 Dec. 25	Riverside County	6.6-7.0	VI-VII
1915 Jun. 23	Imperial County	6.3	V
1918 Apr. 21	Riverside County	6.8	VI-VII
1923 Jul. 23	Riverside County	6.2	V
1937 Mar. 25	North of Borrego Spgs	5.9	
1940 May 18	Imperial County	6.7	V
1942 Oct. 21	Near Borrego Valley	6.3-6.5	V-VI
1954 Mar. 19	Santa Rosa Mts	6.2	
1968 Apr. 9	Ocotillo Wells	6.4-6.8	VII
1969 Apr. 28	Near Borrego Valley	5.8	
1979 Oct. 15	Imperial County	6.5-6.6	
1987 Nov. 23	Superstition Hills	6.2	
1987 Nov. 24	Superstition Hills	6.6	

Future earthquake estimates. McEuen and Pinckney (1972) estimated a maximum probable earthquake for the San Jacinto fault as a whole of magnitude 6.9 to 7.3, and a maximum credible quake of magnitude 7.6. Assuming earthquake rupture of the entire length of the fault, Woodward-Clyde Consultants (1986) similarly suggested a maximum earthquake magnitude range of 7 to 7.2.

Wesnousky (1986) suggested expected earthquake magnitudes between 6.4 and 6.6 for each of the six segments from the Anza gap in Riverside County southward to the Superstition Hills in Imperial County. Anderson, Agnew, and Rockwell (1989) estimated M6.5 as the maximum plausible earthquake for each of the four adjoining 20- to 30-mile-long segments from north of Anza to the Superstition Mountain segment in Imperial County, but they also suggested the possibility of multiple-segment quakes up to magnitude 7.2. Rockwell also has proposed that the Anza gap segment may be capable even of a magnitude 7.0 quake, as there has been no large earthquake there at least since 1892. This whole range of estimates is consistent with the 100-year historic record, in which all 13 earthquakes have fallen in a magnitude range of 5.7 to 7.0.

As the Anza gap, Coyote Creek, and Borrego Mountain segments are only 60 miles from San Diego, earthquakes of magnitude 6.0 probably would produce shaking intensities of MM IV to V in the city and VII to VIII in the northeastern corner of the County. Quakes of magnitude 6.5 and 7.0 could shake San Diego at MM V to VI and VI to VII, and northeast County at VIII to IX and IX to X.

SAN ANDREAS FAULT ZONE

The 150-mile-long southernmost segment of the San Andreas fault extends from Cajon Pass to the Salton Sea. While to the north it is a narrow, well-defined boundary between the North American and Pacific plates, in the Salton Trough the San Andreas splays into a broad zone that includes the San Jacinto, Imperial, and other major faults to the south.

While there has been high seismicity on some of these other faults, the San Andreas in this region has not had a major earthquake in historic times. An excavation at Cajon Pass has provided evidence for at least two, and perhaps four, earthquakes from 1290 to 1805 A.D., and there may have been as many as six quakes there in the past 1000 years. Thus this segment has large earthquakes at least once every two to three centuries, and there appear to have been none here during the past 200 years. As the last quake may have occurred here as long ago as December, 1812, this southern segment is a probable source for the next destructive earthquake in southern California.

Woodward-Clyde Consultants (1986) suggested a maximum magnitude of 7.5 for this segment of the San Andreas fault. Anderson, Agnew, and Rockwell (1989) estimated a maximum plausible magnitude of 7.3 for this segment but up to 8.2 for rupture of multiple segments. A magnitude 8 earthquake on the southern segment could produce intensities of VII to IX in eastern San Diego County and VI to VII in the coastal zone.

IX. EARTHQUAKE OUTLOOK FOR SAN DIEGO COUNTY

As certainly as the Pacific plate will continue its northward migration past North America, so will crustal rocks west of the San Andreas, San Jacinto, Elsinore, Rose Canyon, Coronado Bank, San Diego Trough, and San Clemente fault zones go on lunging fitfully past rocks to their east. While the huge swarm of northeast-southwest faults on the southern coast has not been active for more than a hundred thousand years, the Rose Canyon fault itself clearly remains active, as apparently do several of its urban neighbors -- the La Nacion, those crossing Silver Strand, and perhaps the Florida Canyon and Texas Street faults.

The table summarizes from the text the estimates of maximum magnitude and maximum intensity to be expected from future earthquakes on these faults in or near San Diego County. The message from this table reiterates one that has been repeated many times but always for too small an audience:

In the words of Charles Richter in 1959: "There has been a general impression that earthquake risk does not exist at San Diego, historical records to the contrary being forgotten or ignored...The maximum reasonable intensity during future earthquakes is VIII for the low sandy area on which most of the city, including the business center and the harbor, is situated."

From Michael Reichle, Seismologist with the California Division of Mines and Geology, in 1987: "I would say that there is no place in California that should not be prepared for an earthquake of magnitude 6 to 7."

According to John Anderson, seismologist formerly at Scripps Institution of Oceanography, in 1987: "Eventually San Diego will get stronger shaking than any previously recorded. A quake measuring 6.5 to 7 emanating from the Rose Canyon fault would not be a scientific surprise."

MAXIMUM MAGNITUDE AND INTENSITY ESTIMATES
SAN DIEGO COUNTY

FAULT	MAGNITUDE	INTENSITY COAST	EAST COUNTY
San Miguel	6.0		
Agua Blanca	6.5-7.2	VI-VII	
San Clemente	6.6-7.7		
San Diego Trough	6.1-7.7	VI-VII	
Coronado Bank	6.0-7.7	VI-VIII	
Rose Canyon	6.2-7.0	VIII-IX	
La Nacion	6.2-6.6		
Elsinore	6.9-7.6	VII-VIII	
segment A	6.8-7.3	IV-VII	VII-X
segment B	6.6-7.1	IV-VII	VII-X
segment C	6.8-7.2	IV-VII	VII-X
segment D	6.5-7.2	IV-VII	VII-X
Laguna Salada	6.5-7.0	VII-X	
San Jacinto zone			
Casa Loma-Clark	7.1	VI-VII	IX-X
Anza gap	6.4-6.6	V-VII	VIII-X
Coyote Creek	6.4-6.6	V-VII	VIII-X
Borrego Mountain	6.4-6.8	V-VII	VIII-X
Superstition Mntn.	6.4	IV-VI	VII-IX
Superstition Hills	6.4-6.7	IV-VI	VII-IX
multiple segments	6.9-7.3	V-VII	VIII-X
San Andreas	7.3-8.2	VI-VII	VII-X

X. GOVERNMENT OFFICIALS, COMMUNITY LEADERS, AND INDIVIDUALS CAN MINIMIZE DAMAGE AND INJURY BY PREPARING FOR FUTURE EARTHQUAKES

The lack of strong, damaging earthquakes in San Diego County during this century has been a mixed blessing. We have been spared the death and destruction that have been visited upon San Francisco, Long Beach, San Fernando, and other California cities during that time, but as a result the residents of San Diego County have been misled into a sense of complacency regarding earthquake hazard here. The majority today seem to assume that this part of California has not, does not, and will not have damaging earthquakes, though the older historic record and the geologic evidence show clearly that this is not the case. San Diego County has repeatedly suffered earthquakes of magnitude 6 and higher and shaking intensities as high as MM VII, and similar earthquakes are certain to occur in the future.

Only in the past few years has a concerted effort been made to educate County residents to the earthquake hazard and to encourage preparedness measures. The beginnings of this effort came in 1975, when a county committee, which was convened as a result of the 1971 San Fernando earthquake, issued its report as the Seismic Safety Element of the San Diego County General Plan (San Diego County Environmental Development Agency, 1975). Various earthquake awareness activities were held during statewide Earthquake Awareness Weeks each April through 1986.

In April, 1987, this program was expanded, as San Diego County led the way in developing more extensive informational programs in the state's first Earthquake Awareness Month. That program, and the two annual ones that have followed it, have been organized by Jan Decker, Director Dan Eberle, and their fellow officials at the San Diego County Office of Disaster Preparedness. Meanwhile Supervisor Susan Golding had formed the San Diego County

Earthquake Preparedness Committee, which, in addition to its on-going task of making recommendations for public policy on earthquake preparedness issues, assumed sponsorship of Earthquake Awareness Month activities in 1988 and 1989. During each of these first three Awareness Months there have been conferences for business, government, and community leaders as well as a variety of public lectures and field trips; workshops for hospitals, schools, and businesses; and dissemination of information on earthquake hazard and preparedness measures. Watch for Earthquake Awareness Month events each April.

In the meantime what can you do to prepare for earthquakes to come? Among many sources of information, probably the readiest to hand is the Pacific Bell telephone directory, which has four pages (B9 to B12 in the 1989 edition) of its survival guide devoted to earthquake preparedness. The Sunset book Earthquake Country (Iacopi, 1972), which also is a very good source of information on earthquakes in general, has a section on earthquake preparedness, and in March, 1982, Sunset magazine also printed a special report on earthquakes. Additional information can be obtained from the San Diego County Office of Disaster Preparedness at 5201 Ruffin Road in San Diego (telephone 565-3490).

An earthquake hazard exhibit was prepared for display at the San Diego State University Department of Geological Sciences booth in the Gems and Minerals Building at the Del Mar Fair in 1989 and perhaps in future years. Part of this exhibit is a model of the Rose Canyon fault zone between Mission Valley and La Jolla. This model was constructed by SDSU geology student Bruce Bailey and was made possible by grants from the San Diego and Mission Valley Exchange Clubs, thanks to member Dave Dunn. The model will be part of a permanent earthquake hazard display in the Tecolote Park Natural Reserve Visitor Center, which is planned for completion in 1990. The visitor center and its use for earthquake awareness

education will be the reward of years of effort by Karl
Anderson and fellow members of the Tecolote Canyon
Citizens Advisory Committee, with the assistance of San
Diego City Park and Recreation officials -- especially Pete
Jungers, Lisa Marquez, and Don Prisby, and of Councilman
Bruce Henderson.

Following is a list of some of the important measures that
should be taken by individuals, families, and people in
businesses and offices in preparation for and in response to
earthquakes:

BEFORE THE QUAKE

Develop an earthquake plan for your home or office.
Prepare yourself, your family, and your employees by
completing the activities on this checklist. Businesses and
institutions should choose leaders to be responsible for
preparedness and response measures.

Decide how and where your family will reunite if
separated, and urge employees to have family reunification
plans.

Know the safe spots throughout your home and business --
under sturdy tables, desks, or doorways; against inside
walls; etc.

Know the danger spots -- windows, mirrors, hanging
objects, fireplaces, and unsecured furniture, cabinets,
shelves, etc.

Plan evacuation routes. Conduct practice drills.
Physically place yourself in safe locations.

Learn first aid and CPR.

Keep a list of emergency phone numbers.

Learn how to shut off gas, water, and electricity in case the lines are damaged. Keep wrenches handy for turning off gas and water.

Check all areas of home and business for hazards. Check chimneys, roofs, walls, foundations for stability. Make sure your house is bolted to its foundation.

Secure water heater, appliances, and machinery that could move enough to rupture utility lines.

Keep breakables and heavy objects on bottom shelves.

Secure heavy tall furniture, shelves, etc., that can topple.

Secure hanging plants, heavy picture frames or mirrors (especially over beds), and other hanging objects.

Put latches on cabinet doors to keep them closed during shaking.

Keep flammable or hazardous liquids such as paints, pest sprays, or cleaning products either in cabinets or secured on lower shelves.

Maintain emergency food, water, and other supplies, including a flashlight (with extra batteries), a portable battery-operated radio (with extra batteries), medicines, first aid kit, and clothing. Plan for at least a 72-hour supply of everything.

Have fire extinguisher in working order.

DURING THE QUAKE

If indoors, stay there. Get under a desk, table, or doorway, or stand in a corner. Stay away from windows, fireplaces, tall furniture, falling plaster-tiles.

If outdoors, get into an open area away from trees,

buildings, walls, and power lines.

If in a highrise building, stay away from windows and outside walls. Get under a table. Do not use elevators.

If driving, pull over to the side of the road and stop. Avoid overpasses and power lines. Stay in the vehicle until the shaking is over.

If in a crowded public place, do not rush for the doors. Move away from display shelves containing objects that could fall.

AFTER THE QUAKE

Check for injuries. Apply first aid. Do not move seriously injured individuals unless they are in immediate danger.

Do not use the telephone immediately unless there is a serious injury or fire.

Check for gas and water leaks and for broken electrical wiring or sewage lines. If there is damage, turn utility off at the source. Use flashlight; do not use matches or candles unless you are certain that there are no gas leaks. Do not relight gas pilots.

Check building for cracks and damage, including roof, chimneys, and foundation.

Check food and water supplies. Emergency water may be obtained from water heaters, melted ice cubes, toilet tanks (not bowls), and canned vegetables.

Turn on your portable radio for instructions and news reports. Cooperate fully with public safety officials.

Do not use your vehicle unless there is an emergency. Keep the streets clear for emergency vehicles.

Be prepared for aftershocks.

XI. BIBLIOGRAPHY

Anderson, J. G., Agnew, D., and Rockwell, T. K., 1989, Past and possible future earthquakes significant to the San Diego region. Earthquake Spectra.

Bolt, B. A., 1978, Earthquakes, a primer. W. H. Freeman.

Iacopi, R., 1972, Earthquake Country. Lane Books, Menlo Park, 160 p.

Kern, J. P., 1987, Earthquake shaking and fault rupture in San Diego County. Preliminary report to the San Diego County Office of Disaster Preparedness, 58 pages, fault map of San Diego County.

Leighton and Associates, 1983, Seismic safety study for the city of San Diego, 39 p.

McEuen, R. B., and Pinckney, C. J., 1972, Seismic risk in San Diego. San Diego Soc. Nat. Hist. Transactions, v. 17, no. 4, p. 33-62.

Reichle, M. S., and others, Planning scenario for a major earthquake, San Diego-Tijuana metropolitan area, preliminary report, Cal. Div. Mines and Geology.

San Diego County Environmental Development Agency, 1975, Seismic Safety Element, San Diego County General Plan.

Sunset magazine, 1982, Getting ready for a big quake, March, p. 104-111.

Toppozada, T. R., Real, C. R., and Parke, D. L., 1981, Preparation of isoseismal maps and summaries of reported effects for pre-1900 California earthquakes. Cal. Div. Mines and Geology Open-file Report 81-11.

Toppozada, T. R., Parke, D. L., Jensen, L., and Campbell, G., 1982, Areas damaged by California earthquakes, 1900-1949. Cal. Div. Mines and Geology Open-file Report 82-17.

Wesnousky, S. G., 1986, Earthquakes, Quaternary faults, and seismic hazard in California. Journal Geophysical Research, v. 91, no. B12, p. 12,587-12,631.

Woodward-Clyde Consultants, 1986, Evaluation of liquefaction opportunity and liquefaction potential in the San Diego, California urban area. Final technical report for U. S. Geological Survey Contract No. 14-08-0001-20607.